ENVIRONMENTAL ECONOMICS
&
THE MINING INDUSTRY

Studies in Risk and Uncertainty

edited by W. Kip Viscusi
Department of Economics
Duke University
Durham, North Carolina 27706

Previously published books in the series:

Luken, R.: ENVIRONMENTAL REGULATION:
    TECHNOLOGY, AMBIENT AND BENEFITS-
    BASED APPROACHES
Shubik, M.: RISK, ORGANIZATIONS AND
    SOCIETY
Edwards, W.: UTILITY THEORIES: MEASUREMENTS
    AND APPLICATIONS

ENVIRONMENTAL ECONOMICS & THE MINING INDUSTRY

edited by

Wade E. Martin
Environmental Policy Center
Mineral Economics Department
Colorado School of Mines
Golden, Colorado

KLUWER ACADEMIC PUBLISHERS
Boston/Dordrecht/London

Distributors for North America:
Kluwer Academic Publishers
101 Philip Drive
Assinippi Park
Norwell, Massachusetts 02061 USA

Distributors for all other countries:
Kluwer Academic Publishers Group
Distribution Centre
Post Office Box 322
3300 AH Dordrecht, THE NETHERLANDS

Library of Congress Cataloging-in-Publication Data

Environmental economics and the mining industry / edited by Wade E.
  Martin.
          p.   cm. -- (Studies in risk and uncertainty)
      Includes bibliographical references.
      ISBN 0-7923-9404-6 (alk. paper)
      1. Mineral industries--Costs.   2. Mineral industries-
  -Environmental aspects.   3. Liability for environmental damages.
  4. Pollution--Law and legislation--Compliance costs.   I. Martin,
  Wade E.   II. Series.
  HD9506.A2E6   1994
  338.2'3--dc20   93-32888
                                                              CIP

CONTENTS

I.   Wade E. Martin: **Mining and the Environment:
        An Introduction**   1

        Natural Resource Damage Assessment: Valuing
                Nonmarket Goods   4
        Global Warming and the Mining Sector   9

II.   V. Kerry Smith: **Natural Resource Damage Assessments
        and the Mineral Sector: Valuation in the Courts**   15

        Introduction   15
        Natural Resource Damage Assessment: An Overview   17
        Applying Nonmarket Valuation Methods in NRDA   22
        Complete Restoration, Strict Liability, and
                Efficiency   34
        NRDA and the Mineral Industry   35

III.   Ronald G. Cummings, Philip Ganderton & Thomas McGuckin:
        **Valuing Environmental Damages With the Contingent
        Valuation Method: A Critique**   53

        Overview of the Problem   53
        Do CVM Subjects Understand the Good Which
                They are to Value?   54
        Do CVM Subjects Understand the Hypothetical
                Market?   63
        What is the Empirical Content of CVM Value
                Responses?   67
        Concluding Remarks   72

IV.   Ralph C. d'Arge: **Intergenerational Fairness & Global
        Warming**   79

        Introduction   79
        Regional Impacts of Global Warming   80
        Intertemporal Impacts of Global Warming   89
        Economic Mitigation Strategies   93
        Summary and Conclusions   97

V.   H. Stuart Burness and Wade E. Martin: **The Effects of**
       Global Warming on the Mining Industry: Issues,
       Tradeoffs & Options   107

       Introduction   107
       The Physical Nature of Global Warming   109
       A Heuristic Economic Model of Global Warming   112
       Effects on the Mining Industry   119
       Summary and Conclusions   124

CHAPTER
I

MINING AND THE ENVIRONMENT: AN INTRODUCTION

Wade E. Martin

The four core chapters of this book (chapters 2 through 5) are based upon the 1990 John M. Olin Distinguished Lectureship Series on Mining and the Environment held at the Colorado School of Mines. This topic was chosen due to the need to address the critical issue of the impact of environmental concerns, and legislation generated by these concerns, on the mining industry. The environmental concerns are as diverse as the mining industry itself. It is important to realize that the mining industry, by its very nature, impacts the environment and society must explicitly consider the tradeoffs involved with extracting minerals, pumping oil or drilling for natural gas.[1]

The diversity of the environmental concerns range from the legislative mandates of the United States Congress to the "grassroots" movements at local mining sites. The legislative concerns originate at all levels of government, from the federal down to local governmental units. The legal environment affecting a mining operation can change dramatically from site to site due to the variety and complexity of environmental concerns at the local level. The legal approach pursued in southeastern Alaska is quite different than that pursued in rural Missouri. Consistency and uniformity at the federal level is also difficult to achieve.

The federal impacts on the mining industry from environmental legislation are quite significant. One of the earliest federal environmental actions affecting mining was the National Environmental Protection Act of 1969 (NEPA). The major feature of this law for the mining sector is the requirement that an Environmental Impact Statement (EIS) or an Environmental Assessment (EA) be prepared prior to developing a mine, exploring for oil, or any similar activities on federal land or that would affect federal land.[2] The EIS process can add an average of one to two years to the planning stages of a project as well as cost millions of dollars.

Once NEPA was passed in 1969, legislative action at the federal level increased dramatically over the next two decades. The legislative action on environmental issues over the following decades included the Clean Air Act (CAA) of 1970 (and the amendments of 1977 & 1990), the Federal Water Pollution Control Act (Clean Water Act, CWA) of 1972 (and the amendments of 1977 & 1987), the Safe Drinking Water Act of 1974 (amended in 1984), the Trans-Alaska Pipeline Authorization Act of 1973, the Endangered Species Protection Act of 1973, the Surface Mining Control & Reclamation Act (SMCRA) of 1976, the Resource Conservation & Recovery Act (RCRA) of 1976, the Toxic Substances Control Act (TSCA) of 1976, the Comprehensive Environmental Response, Compensation, & Liability Act (CERCLA) of 1980, the Superfund Amendments & Reauthorization Act (SARA) of 1986,[3] and most recently the Oil Spill Prevention Act of 1990. Although this is not a complete list of all the environmental legislation during the relevant timeframe, it quickly becomes evident that impacts on the mining industry can result from a wide range of legislative sources.

To address the impact of each of these acts on the mining sector is well beyond the scope of this book. Therefore, this series of lectures was designed to provide insight into two important issues facing the mining sector at this time. The first issue concerns the liability associated with CERCLA damages (or

Superfund liability). This is an issue that is currently having an dramatic impact on the industry. The impact is being felt in transactions involving the potential sale of properties, insuring operations, development of new properties, joint ventures, or more generally, practically every phase of the mining firms operation.

The second issue focuses on an environmental topic that has not been specifically addressed in federal legislation, although it has been indirectly considered, that is global warming or the "greenhouse effect". One of the interesting aspects to this environmental problem is the uncertainty associated with it at every phase of the analysis. The predictions of the general circulation models of climatologists are questioned due to the uncertainty of ocean effects, urbanization, etc. (see Burness & Martin, Chapter 5). The economic models are criticized for the uncertainty associated with the benefit estimates from reducing greenhouse gases, particularly carbon dioxide (CO_2), concentrations in the atmosphere as well as estimates of the cost of reducing GHG concentrations and/or emissions. This raises the interesting question of what is the optimal policy and what will be the impact of this policy(s) on the mining sector, given the uncertainty.

The first of these two topics is addressed by V. Kerry Smith and Ronald G. Cummings, *et al.* Professors Smith and Cummings were chosen due to their pioneering work in the area of valuation of nonmarket goods, particularly involving the use of survey methods. The issue of global warming is addressed in two papers; the first by Ralph C. d'Arge the second by H. Stuart Burness and Wade E. Martin. The international experience of Professor d'Arge regarding equitable distribution schemes makes him well qualified to address the critical issue of intergenerational "fairness" and global warming. Professors Burness and Martin combine theoretical and policy experience to discuss the background issues necessary for understanding the uncertainties involved and the impacts of the various policy options.

The rest of this introduction will highlight the findings of these authors.

The link to the mining industry is emphasized much less in Cummings, *et al.* and d'Arge. These two chapters, however, are important to understanding the impact on the mining industry from the two sources of environmental regulation considered in this book. Each of the chapters is based upon the unique experiences of the authors and provides their perspectives on the important issues associated with environmental impacts and issues on the mining sector.

NATURAL RESOURCE DAMAGE ASSESSMENT: VALUING NONMARKET GOODS

Natural resource damage assessment is designed to compensate the public for reductions in the value of natural assets that occurs from the actions of others. For example, the compensation required from Exxon for damage to the natural environment resulting from the oil spilled from the Exxon Valdez. These types of damages do not include liability for "humans' economic losses" (see Smith, Chapter 2), for example, the losses of commercial fishermen.[4] Estimating the reduction in the value of these natural assets is not a straightforward task. Smith provides an excellent foundation for understanding many of the issues involved, both for the novice and those already familiar with the topic through his discussion in the text and his methodological review in Appendix A to Chapter 2. Cummings *et al.* then addresses some of the more problematic shortcomings of the existing survey methodology and its acceptance as the "best available technology" for natural resource damage assessment.

The legislative history of natural resource damage assessment has its origin in the Trans-Alaska Pipeline Authorization Act of 1973. As Smith points out, this act provides for the public trustee to sue for damages to the natural environment when an oil spill occurs. The Superfund legislation (CERCLA & SARA) has extended the natural resource damage liability to the case of hazardous substances. It is this extension that provides the greatest impact on the mining sector. Prior to the Superfund legislation, wastes generated by the mining sector were explicitly excluded from federal environmental actions. For example, mining wastes are considered to be "high volume, low level hazardous wastes" and are excluded from regulation under RCRA by the "Bevill" amendments (see Smith, Chapter 2).[5]

To understand the role of the public trustee in recovering damages to natural assets from actions by others it is necessary to understand what is meant by the terms "economic value" and a "nonmarket good". Economists define goods as either market or nonmarket goods. A market good is sold in an organized market where the price at which the exchange takes place is known and generally determined by market forces. A nonmarket good, however, is defined as a good that does not trade in an organized market and there is no explicitly stated price for the good. Examples of market goods are copper, cars, and opera tickets, whereas, examples of nonmarket goods are clean air, reduced cancer deaths, and increased wildlife habitat.

Once we realize that there are nonmarket goods, what Smith calls natural assets and the services they provide, we must then determine the value of these assets. The economics literature identifies two types of values: use values and nonuse values. An example of the use value of a natural asset (nonmarket good) would be hiking in a wilderness area. Nonuse values have been divided into two primary categories: option values and existence values. The option value associated with a nonmarket good is the value an individual would place on preserving the asset for his or her possible use in the future. For example, if someone living in New York would be willing to pay to preserve the Grand Canyon, even though they have never visited the site, because they plan to visit the site at some future date then the amount they are willing to pay is their option value for this natural asset.

Existence value is similar to option value, except that a person is willing to pay a given amount because they derive utility from the fact that the natural asset exists, even though they have no intention of visiting the site in the future or ever using the site. Both option and existence values (the nonuse values) are quite difficult to estimate as both Smith and Cummings point out. However, it is important to realize that nonuse values, as well as the use values, must be estimated based upon the ruling by the D.C. Circuit Court of Appeals in their evaluation of the guidelines set forth by the Department of Interior for natural resource damage assessment. Smith and Cummings *et al.* discuss the DOI guidelines and the ruling by the Washington D.C. Circuit Court of Appeals in 1989 that confirmed the relevance of nonuse values in the natural resource damage assessment process.

Smith discusses the methodologies available for estimating both use and nonuse values of natural assets and discusses their impact on the mining industry. Smith highlights his discussion of the natural asset valuation process with an analysis of the Eagle River case in Colorado. An interesting finding of this analysis is that the estimates of the "average" value per person for the natural asset (the Eagle River) where similar for both the plaintiff and defendant in the case, however, the total value that the public trustee (State of Colorado) sued for was 84 times the estimate of the defendant (Gulf-Western Industries) in the case. Smith attributes this difference to the issue of "defining the extent of the market" for the natural asset.

Smith concludes his presentation with a discussion of natural resource damage assessment and the mineral industry. He points out that the impact of the doctrine of joint and several liability specified in the Superfund legislation will be felt throughout the mining sector, particularly with respect to property appraisals and the long-term liability position of firms that may no longer be involved in the mining industry, let alone a specific mine. Also of importance to the mineral industry is the growing list of "public trustees". Initially, public trustees were confined to federal government representatives, however, this has been extended to include all levels of government and Indian tribes. This extension results in approximately 69,000 sources of possible law suits.

Cummings *et al.* focus on the use of survey techniques, the contingent valuation methodology (CVM), to estimate use and nonuse values from natural resource damages. They find fault with the circuit courts ruling that the CVM is an appropriate methodology for estimating damages. They base their criticism of the courts acceptance of CVM on three basic issues: do respondents understand the "nonmarket good" or natural asset they are asked to value, do the respondents understand the "hypothetical market" structure used, and the empirical content of the CVM value.

The focus of the discussion regarding the first of these issues by Cummings *et al.* is the relationship between the theoretical basis of traditional value theory in economics and the use of values derived from a "hypothetical" transaction. The basic question to be answered is "Is the theory of consumer behavior under conditions where both transactions and payments are real applicable to behavior . . . where both transactions and payments are hypothetical?" To answer this question Cummings *et al.* provide a discussion of the microeconomic foundation for "real markets" and then consider the empirical evidence from the "hypothetical markets" for support of the survey method.

The review of the theory of value stresses the importance of understanding the good to be valued and how this good enters into an individuals choice set. The importance of understanding the good derives from the necessity to make trade-offs within the choice set. When asked to value a good, an individual must re-evaluate all the goods in the choice set to determine the preference for the good being considered. If there is not a clear understanding a particular good then the resulting valuation may not be correct.

The conclusion of Cummings *et al.* is that their interpretation of the empirical evidence suggests survey respondents do not understand the good to be valued and, therefore, cannot provide a meaningful answer to the valuation question for a nonmarket good. The primary reason given for their conclusions is that respondents too easily change their valuation when provided with what should be theoretically benign changes in the context in which they are to provide their valuation information.

The second concern of Cummings *et al.* focuses on the respondents understanding of the "hypothetical market" structure that is used in the valuation process. The importance of this concern is the link between the respondents stated valuation and the "real" economic commitment that it reflects. In effect, if we "ask a hypothetical question are we going to get a hypothetical answer" that does not truly reflect a real commitment by the respondent!

Once again, the authors turn to the empirical evidence in an attempt to answer this question. The empirical studies they cite lead them to the conclusion that respondents do not fully understand the commitment involved in their response to the valuation questions in a survey format.

The final question asked by Cummings *et al.* is more technical in nature, basically, "what is the empirical content of the responses?" The data that are obtained from a contingent valuation survey are then used for empirically calculating willingness-to-pay or willingness-to-accept values for the sample and then applied to the population. The main concern of the authors is the generally accepted practice of "removing" various data points based upon the belief that they represent either protest bids or are outliers in some, perhaps arbitrary, sense.

Based upon the authors analysis of the various methods used to address these two critical issues, their conclusion is that the "current state of the art involves little more than arbitrary choices for what the researcher takes to be a reasonable or an unreasonable value". The importance of "judgement" in the empirical analysis of the data by the contingent valuation researcher provides an important source of concern regarding the stated values used for decision-making in allocating scarce resources.

The fundamental conclusion of Cummings *et al.* is that the empirical evidence to date does not support the use of the contingent valuation methodology as the "best available technology" for valuation purposes under the natural resource damage assessment requirements of Superfund.

Smith and Cummings *et al.* address an issue that is currently having a great impact on the mining industry. The complexities of the natural resource damage assessment process is difficult to grasp since the federal government (through the Department of the Interior) and the judicial system are currently involved in defining the issues. These two chapters, however, should help in understanding many of these complexities.

GLOBAL WARMING AND THE MINING SECTOR

There are many dimensions to the global warming issue. The physical aspects of the greenhouse effect are extremely complex with the general circulation models (GCM) of climatologists typically solving 200,000 equations on each run (see d'Arge, chapter 4). However, even with this degree of sophistication the GCMs are often criticized for their lack of precision in predicting temperature changes although the models generally agree on the magnitude of the change. The GCMs have consistently predicted a change in the mean temperature of 1.5° to 5°C when CO_2 concentrations double from the base year concentrations in 1860. As Burness and Martin note, (chapter 5) the critics of the climate models generally focus on the shortcomings of the models but have not disproved them. Therefore, the important issue for the mining sector becomes determining the impact on the industry given the various policy options that may be implemented. Such analysis would also be useful in determining the direction that the policy maker should pursue.

Burness and Martin focus on the impact of current policy options, whereas, d'Arge addresses the issue of intergenerational "fairness" or the distributional effects associated with global climate change. d'Arge considers three issues: the regional impacts of global warming; the intertemporal impacts of global warming; and an economic mitigation strategy involving intergenerational transfers.

The first of these issues involves a discussion of the regional impacts of the expected climate change and the identification of the regions that will benefit from the change and those that will lose. Realizing that the GCMs estimate a change in the mean temperature of the earth of 1.5° to 5°C d'Arge focuses on the distribution of this change since not all areas will be equally affected. The main conclusion of this section is that the countries that are currently wealthy will be able to adapt, whereas, those that are poor will be severely affected by a climate change, in the absence of any preventive action. Therefore, as d'Arge points out "the rich will get richer at the expense of the poor".

The second issue d'Arge considers is the extent and speed with which global warming will occur and the impact that this will have on the economies of the affected countries. Due to the past use of fossil fuels the earth is already committed to some global warming and continued use of these energy sources will result in even more warming. The importance of the additional warming is whether or not the ecosystems will have sufficient time to adapt to the temperature change and to migrate as necessary. If warming occurs too rapidly, then the near term costs could be significant, whereas, if the warming occurs gradually then the near term benefits, such as enhanced plant growth due to CO_2 fertilization effects, may dominate. Without specifying a particular policy, d'Arge does not speculate as to which scenario is most likely.

d'Arge's final section presents an interesting paradigm for intergenerational transfers to address the distributional problems that arise with global warming. The transfer mechanism that d'Arge suggests is one where the current "rich" nations would transfer wealth to the "poor" nations. According to d'Arge, such a transfer mechanism should improve global intertemporal efficiency. This raises the interesting question as to why would the rich nations agree to such a transfer arrangement?   To answer this question d'Arge references the work of Kahneman and Knetsch who suggest that individuals derive utility from knowing that the "right thing is being done". d'Arge expands on this transfer paradigm in his appendix.

Burness and Martin focus on the policy aspects of global climate change and begin their analysis with a discussion of the physical aspects of global warming. This discussion is followed by a brief presentation of the economics of common property resources and how global warming fits into this category of resource problems. The use of the atmosphere as a sink for CO_2 emissions has been treated as a "free" good to each firm and/or country, therefore, encouraging overuse of the resource. Using this as a foundation, Burness and Martin present a heuristic economic model of global warming as a common property resource based upon the earlier work of Nordhaus. The model is based upon the traditional benefit cost analysis of welfare economics where the focus is on marginal benefits of reducing greenhouse gases and marginal abatement costs to determine the appropriate level of action.

Using this framework, Burness and Martin evaluate the various policy impacts of global warming on the mining sector. The basic conclusion that they arrive at is that the direct effects on the mining sector of global warming would be minimal, however, the indirect effects could be quite significant. The critical features in determining the impact on the mining sector are the policy option(s) pursued by the various governmental entities and the time frame considered, i.e., near term versus long term effects.

To determine the "optimal" policy to address the greenhouse warming is difficult due to the high degree of uncertainty. A related issue that also generates considerable uncertainty is that if an optimal policy is identified, will the government implement such a policy. However, to at least consider the possibilities, several studies were reviewed and the impact of the various policy options were analyzed in the context of their impact on the mining sector. The main conclusion of Burness and Martin is that a comprehensive policy to address global warming will face major complications that arise from the joint problem of regional variation of greenhouse effects, as shown in d'Arge, and the multiple political objectives that must be considered and coordinated. Therefore, pursuing a policy that involves implementing low cost strategies, such as eliminating chlorofluorocarbons, may be the appropriate action to take until some of the uncertainty both with the GCMs and the economic analyses are resolved. If this course of action is pursued, the near term impact on the fossil fuel sector of the mining industry should see minimal direct effects.

The chapters in this volume address only two of the many

environmental issues confronting the mining sector. These two issues, however, encompass many of the important aspects of the interaction between the desire to create and maintain a given level of environmental quality and to continue to produce the goods we use to make our everyday life more enjoyable.

ENDNOTES

1.   The definition of mining for our purposes will be quite broad. We will include all aspects of the mining process, e.g. hard rock mining, fossil fuel extraction, etc.

2.   For a more extensive discussion of the EIS/EA process see Martin & Winters, 1991.

3.   CERCLA and SARA together represent the so-called Superfund legislation.

4.   Economic losses to humans are already covered by conventional tort law since these are market losses.

5.   This exclusion is currently being studied by the Environmental Protection Agency. EPA has developed guidelines for regulating mining wastes that have been distributed for comment, but have not yet developed a Proposed Rule.

CHAPTER
II

NATURAL RESOURCE DAMAGE ASSESSMENTS AND
THE MINERAL SECTOR: VALUATION IN THE COURTS

V. Kerry Smith*

INTRODUCTION

The instability of oil prices following the latest Middle East conflict provides a clear reminder of how politically induced episodes of resource scarcity (paralleling the oil embargo some fifteen years ago) can significantly impact aggregate economic activity around the world. Now it may seem that the environmental concerns that have occupied the "front burner" of most countries' policy agendas will be displaced for the foreseeable future by the more immediate issue of price stability through supply augmentation. With their displacement of environmental concerns, activities in the extractive and especially the energy related sectors of our economy can return to "business as usual."[1] In what follows, I will argue that this conclusion is not only short-sighted, it is inconsistent with the existing constraints on the activities of extractive industries that were defined <u>before</u> the 1973-74 oil embargo and progressively expanded in the intervening years. These restrictions were designed to phase into effect with the environmental legislation of the late 1970s and early 1980s. Now we are beginning to see their effects. Moreover, one of the most important of these influences traces its origins to the Trans-Alaska

Pipeline Authorization Act passed nearly two decades ago. This legislation introduced a new instrument for resource and environmental policy--natural resource damage liability.

Natural resource damage liability implies that firms involved with hazardous waste or oil must consider the potential impact of these substances on natural resources. This liability treats the natural environment as assets and recognizes that the services they provide outside markets are valuable to society. In effect, it defines a set of property rights for society that are overseen by designated trustees for each type of resource.

Three aspects of natural resource liability are especially relevant to the operations of firms within the extractive sector. First, natural resource damage liability is the only federal constraint currently imposed on residuals produced by firms involved in mineral production and oil exploration and extraction at their initial stages of production. In most other production activities, these residuals would be classified as hazardous wastes. The "Bevill" exclusion to the Resources Conservation and Recovery Act of 1976 (RCRA) exempted "high volume, low hazard" mining wastes (and drilling muds) from treatment under Subtitle C hazardous waste regulation until the United States Environmental Protection Agency (EPA) completed a report to Congress assessing them. This process is still underway.[2] Thus, the only federal role in influencing the residual disposal practices of extractive firms is through the enforcement of liability for damages resulting from releases that have injured natural resources.

Second, the theory and practice of nonmarket valuation techniques are no longer the domain of economists and policy makers assessing public investment decisions or regulatory standards. Now these methods have direct relevance to private firms' liabilities for damages that will be determined through litigation.

Finally, because the incentives for firms' future behavior are different, we can expect that the models economists use to describe them will change. New frameworks must incorporate a recognition of how two types of assets--the private stocks of mineral deposits and the quality and character of the public amenity services--affect their intertemporal extraction profiles.

Since liability for natural resource damages is the only federal influence on the disposal of residuals from mining activities, it is important to understand the source of that liability and the factors influencing its implementation before we can evaluate its effects on the mineral sector. The next section two begins this process by describing the evolution of natural resource damage liability and the economic interpretations that can be offered for its specific requirements. Following that background, section three describes how both current litigation and the economic frameworks for defining the value of nonmarket environmental services affect the estimates produced by those frameworks.

The fourth section discusses the efficient use of natural resource damages for restoration and the incentives that are likely to be created by the process of imposing the liability. The last section conjectures about the future impact of natural resource damage liability for the mineral industry.

NATURAL RESOURCE DAMAGE ASSESSMENT: AN OVERVIEW

Background

Natural resource damage assessments (NRDA) are usually viewed as activities mandated by the Comprehensive Environmental Response, Compensation and Liability Act of 1980 (CERCLA) and its reauthorizing legislation, the Superfund Amendments and Reauthorization Act of 1986 (SARA). Actually, the NRDA story begins in 1973 with the Trans-Alaska Pipeline Authorization Act. The language in that legislation identified strict liability for damages from oil spills but provided only a vague course of action for seeking damages and determining the amount of this liability that would extend beyond cleanup costs and "humans' economic losses."[3] The Deepwater Port Act of 1974 explicitly allowed recovery of damages to the environment itself from oil spills, but not until CERCLA's mandate in 1980 was this liability extended beyond the oil industry and a process defined for measuring damages.

This process was to begin with the President issuing "Type A" regulations for simplified assessments involving small incidents and "Type B" regulations for the major cases. These regulations were to define procedures for damage assessment that would provide their user (the relevant trustee) rebuttable presumption in any court case involving natural resource damage liability.

It has taken nearly a decade (and a major oil spill in Alaska) for the full impact of these requirements to be "noticed," partly because of the delay and confusion associated with implementing CERCLA's legislative mandates. In 1981, President Reagan delegated the responsibility for the regulations to the Department of Interior (DOI), but it took five years and a court ruling before proposed regulations were issued in December 1985 for the Type B cases (and May 1986 for the Type A). Reauthorization of CERCLA and associated amendments to the regulations, together with nine lawsuits questioning the proposed Type B regulations, further delayed the process.[4]

Today, there are a set of rules for Type B assessments that have been partially upheld by a recent (1989) D. C. Circuit Court of Appeals ruling and partially remanded to DOI for revision.[5] While NRDA cases are proceeding, current and future defendants, as well as state and federal trustees, are anxiously awaiting the next round of rules for Type B cases from DOI.

Alternative Economic Perspectives on NRDA

It is remarkably easy to provide an economic rationale for most of the current mandates that appear to define how natural resource damages are to be measured. This section will develop such rationale by explaining the key implications of DOI's Rules and the July 1989 D.C. Circuit Court of Appeals ruling. First, however, it is important to distinguish how economists would analyze each of three distinct aspects of the legislation--its mandates for natural resource damage liability, guidelines for implementation, and incentives for future private activities involving hazardous waste and oil. Economic analyses of these three aspects of NRDA's impact have been confused in past commentaries (see Cicchetti, 1989, and Polinsky and Shavell, 1989).

Economic analyses of legislative mandates accepts the stated intentions and evaluates whether the procedures to achieve the mandate can be "rationalized" given those objectives. For example, by establishing natural resource damage liability, the law requires that natural assets held in public trust be maintained free from the effects of hazardous waste or oil.[6] Moreover, to the extent current or past activities involving hazardous waste or oil lead to releases of substances injurious to natural resources, then those resources must be restored. If restoration is infeasible or the costs would be excessive ("grossly disproportionate") in relationship to the total value of the resources in their pre-release condition, then those parties held responsible must compensate for the lost monetary value. In economic terms, this would include losses incurred in the past because a release affected the flow of values generated by the resource (as an asset) and changes in its total value, thereby reflecting the implications of the injury for the future values that the resource would have generated. Economic rationalization of this mandate amounts to evaluating whether the initial DOI rules and the modifications implied by the court of appeals decision provide appropriate methods for measuring losses due as compensation to "the public." Economic analysis of implementation of the legislative mandates would consider what should be done with the compensation. In effect, how do we determine efficient restoration, and what implications would incomplete restoration have for the amount of the natural resource damages?

By establishing liability for the damages associated with injuries to natural resources from hazardous waste or oil releases, the law will affect the behavior of those involved with these substances. We can evaluate these incentives from at least two perspectives. The first is a normative economic question and the second a positive one. If economists were designing a set of policies to deal with releases of hazardous wastes or oil, would this type of liability rule induce efficient responses on the part of those who deal with such substances? In effect, does this liability rule provide the appropriate incentive in the use of natural assets. Alternatively, in addressing the positive economic issue, we accept the given liability rule and ask about its likely effects on current and future activities involving hazardous waste and oil. This latter issue is considered in section four.

Type B Damage Assessment

Once the relevant trustee has been notified a release has the potential for natural resource injuries, a pre-assessment screening is undertaken to determine whether a full assessment is warranted. If so, as part of the assessment plan, the trustee must determine whether the situation requires a Type A (simplified) or Type B (major) evaluation. With a Type B assessment, each case is assumed to be sufficiently important to warrant individual attention. Scientific criteria are used to provide the basis for defining whether or not the release has injured the natural resource. The DOI protocol for damage assessment requires that injury be linked to the substances released (if not in actual practice, in principle) before any monetary assessment of the damages can take place.[7]

When these conditions are satisfied, the damage assessment involves developing values for any past losses arising from reductions in the flow of services provided by the damaged resource and evaluating what is due the trustee because the resource remains injured at the time of the assessment. The mechanisms for responding to these requirements have been the focus of the DOI rules as well as two court decisions. Table 1 summarizes the key elements in the assessment process that were affected by the court's rulings.

Economic measures of past and current damages require specifying how the scientifically defined injuries affect the services the resource has provided and could provide (in the future) to people. While initial DOI rules limited consideration to what might be described as in situ use, the circuit court ruling allows nonuse values to be included, describing them as a form of nonconsumptive use.[8] No direct market transactions occur in the in situ uses and, of course, "nonuse services" do not leave a behavioral trail in market

Table 1:   Type B Damage Assessment: DOI, Court of Appeals, and Achushnet River/New Bedford Harbor Decisions

Issue	DOI Final Rules for Type B (August 1, 1986)	DC Circuit Court of Appeals Ruling (July 14, 1989)	Achushnet River/New Bedford Harbor Ruling (June 7, 1989)
Timing of Past Damages	Vague; only specific direction is for releases initiated and completed before CERCLA because they are not subject to NRD liability.		For divisible damages, established date of passage of CERCLA (1980) for releases with continuing effects; for indivisible damages, full amount of damage is to be used.
Hierarchy of Valuation Methods	Established hierarchy with "prices" for natural resources.	No hierarchy of methods intended; recognizes that nonmarket valuation is likely to be only feasible basis. Affirms CVM as a "best available methodology."	
"Lesser of Rule" DOI rules required 10 percent real rate in calculating present value of future losses. The Court of Appeals ruling maintained this requirement.	Natural resource damages associated with the current and future status of injured resource should be measured a the lesser of the restoration cost, replacement cost, discounted use values lost as a result of the injury to the resource.	Rejected "lesser of" rule; damages correspond to restoration cost unless restoration is technically infeasible or cost is grossly disproportionate in relation to discounted total value (including use and nonuse) lost as a result of injury. In these cases, damages would be that discounted value.	
Committed Use	Defined as either a current public use or a planned public use for which there is documented legal, administrative, budgetary, or financial commitment before the release.	Removed restriction on nonuse values; defined committed use as requirement to avoid highly speculative future uses.	

choices.[9] Due to this limitation, the methods of nonmarket valuation developed for benefit-cost analyses of environmental policies have been a key focus in applying economics to damage assessment.

Given that values can be measured, the basic structure of an amended DOI process would be consistent with a compensation principle. That is, if designated natural resources are held in public trust as assets and if private actions reduce their value, then the appropriate compensation is that reduction in value. The court of appeals ruling goes beyond this economic interpretation by implying that using a monetary value to compensate for a public asset may consistently understate its social value.[10] Hence, the ruling favors cost of restoration and restoration of the resource to baseline (pre-release) conditions over an economic valuation of what would be lost in the future, unless restoration would be impossible or it imposes costs completely inconsistent with the resource's value.[11]

The appropriate measure of compensation, from an economic perspective, would be the valuation of what would be lost (considering all components of that value). Going beyond that introduces biases, presumably because Congress prefers to err on the side of over-compensation. This may seem to be a conservative position, but it need not be. Generally, the outcome depends on the opportunity cost of the excess compensation: who gives up what to produce the excess compensation, and does that cost exceed any likely economic measure of the damages arising from the release and the associated injury to a natural resource?

APPLYING NONMARKET VALUATION METHODS IN NATURAL RESOURCE DAMAGE CASES

Background

The circuit court of appeals recognized what environmental economists have known for over forty years. Transactions involving the amenity services of natural resources do not take place in organized markets. Equally important, no active markets exist for the natural assets that provide such services. These "natural assets" are not like private assets where measures of loss can rely on a market price. Thus, DOI's original preferred methods for measuring the economic magnitude of the damages were irrelevant to the problem.[12] However, this does not mean that people do not make choices to acquire the *in situ* services of natural resources. They do allocate some part of their incomes (and time) to that process. While these choices can be described as transactions, the information conveyed by these actions is more diffuse than for transactions in a market. Information equivalent to market signals must be reconstructed if these choices are to be used in recovering people's values. This detective work is the essence of economic techniques classified as indirect market methods for valuation.[13]

The exchange connects the price paid to the commodity purchased In a market transaction. As a rule, with nonmarket transactions, economists can observe one component of the price/commodity pair. We must use economic models of what motivated individual behavior to reconstruct the information necessary to impute the other element. Three methods for doing this have played a significant role in environmental benefit analysis, and some version of the first two of them have been used in nearly all NRDAs to date. These three methods are the travel cost recreation demand model, the hedonic property value model, and the averting behavior (or household production) model.

The court of appeals ruling recognized that the valuation tasks involved in a damage assessment are usually not routine applications of these methods. Two aspects of valuing natural assets cannot be completely addressed with any of these methods. As assets, natural resources can be expected to provide services over time. Indirect methods estimate the value to individuals of the service flows as they use them in some observable way. For example, the purchase of a home in an area that has a given level of air quality can be observed and the services provided by the natural asset air can be imputed. However, services are available to people in ways that do not require an observable choice. While the literature has classified these as nonuse values and focused its discussion on behavioral motivations that justify such values, these services have certain public goods characteristics.[14]   Specifically, they are nonexclusive and nonrival in consumption. Because people do not have to take action to obtain them, some economists argue that they leave no behavioral trail.[15]

A second consideration may well be more important to NRDA. This is the behavioral response of individuals to uncertainty. As the time horizon considered in evaluating people's choices lengthens, it is more important to incorporate the added value people may place on ensuring that an asset's services are available under specified conditions. This is one response to future uncertainties about how much or even whether they will want to use a resource's services. Obviously, the added value people place on this assurance will depend on their other mitigation opportunities.

Due to these two dimensions, nonuse values and uncertainty, of valuing the losses from injuries to natural resources, economists have focused research on classifying the components of people's full value for the services of a natural resource and on the use of survey, or direct, methods (contingent valuation) in measuring some or all of these components. The remainder of this section will: (a) describe the implications of adopting classification schemes for valuing the services of natural resources for selecting among nonmarket valuation methods; (b) illustrate the limitations in how these methods have been applied to date; and (c) discuss the research issues that result and how they are likely to affect future NRDAs.

Classifying People's Values for Natural Resources[16]

To classify people's values for natural resources requires consideration of the various sources of these values and what affects the valuation measures. The sources of value for an individual derive from both use and nonuse characteristics of the natural asset and the services it provides. The services provided by a natural asset can be classified as 1) a consumptive use value;

2) a nonuse value analogous to a public good; and 3) a quality feature. To measure the damages associated with an injury to a natural resource, the injury must affect one or more of these services.

Focusing on the direct approach since this is the only method for valuing nonuse aspects of a natural resource, the analyst must consider the level of uncertainty associated with the natural resource damage situation and the economic model of behavior that is involved. *Ex post* valuation measures relate to situations where the analyst assumes little uncertainty exits for the typical individual's decisions. *Ex ante* valuation, however, generally assumes that significant uncertainty exists regarding the typical individual's valuation decision.

When significant uncertainty exists it is necessary to consider the probability that a given situation will exist. Therefore, the analyst must consider the "expected" behavior of an
individual. To determine a monetary measure of the value of some change in the services provided by a natural asset requires specifying the value within the context of a model of the valuation decision. Thus, if there is a "determinate" situation individuals will act to maximize their satisfaction, or utility, from the consumption of the services provided by the natural asset as well as the market goods they consume. When uncertainty is present regarding the services provided by the natural asset, individuals will act to maximize their "expected" utility, based upon the relevant probabilities, from the market goods they consume and the services provided by the natural asset.

The formal definitions of value concepts associated with either maximizing utility or expected utility identify four issues. First, use values will mean different things depending on the treatment of uncertainty and the definition of how the release of hazardous waste or oil affects the resource. For both the *ex ante* and the *ex post* approach, the use values can be measured as either originating from a change in access (a price change) or a change in the quality of the resource.[17] However, using this concept of damage assessment implies that the resource is precluded from any use. This may be quite an heroic assumption in many cases!

Second, to the extent the public goods aspect of the natural asset contributes to well-being, the nonuse values will be affected by what we assume this value is in each situation. By focusing on existence value, the literature to this point has diverted attention from some problems that are likely to have the greatest practical importance. Releases of oil or hazardous wastes do not destroy the natural asset, the release changes people's perceptions of what services they receive from the public good. This has many implications for the valuation process. As in the case of use values, changes in the public good or its quality represent analytically different ways of describing the injury to the natural asset. Neither type of change requires complete elimination of the resource.

Equally important, if the release affects both the private and public aspects of the natural asset and is modelled analytically by presenting the asset after the change has occurred and specifying greater costs for gaining access to the equivalent private services as those provided by the asset prior to the release, then any measure of the values of improvements will be incorrect. The evaluation will be conducted at the wrong baseline. It will be evaluated at the post-release conditions versus the pre-release conditions.

The third issue of concern is how individuals will adjust to uncertainty. In the *ex ante* approach to valuation separate identification of the probabilities (risk premia) requires that the model specify how people adjust to the uncertainty they face. For each specification, a different separation of use from risk premia is possible (see Smith, 1987). In practice, these separations (including the conventional concept of option value that compares a constant state independent *ex ante* value, the option price, with the expected value of state dependent *ex post* values) may not be consistent with the circumstances constraining how people actually make these decisions. These conditions are what will be relevant to how they value the service of the injured natural asset, not some idealized view of their behavior.

Finally, the choice of whether to use a willingness-to-pay (WTP) or willingness-to-accept (WTA) format must be considered. The WTP format has been implicitly adopted in most of the cases to date. The legislative mandates seem to imply that WTA (or the presumption that the public had property rights to the baseline conditions of the resource) would be the correct perspective. Use of a WTP orientation presumably arises from an assumption that differences between WTP and WTA may well be small and that survey approaches cannot adequately measure compensation or WTA (see Cummings, Brookshire and Schulze, 1986 and Schulze, 1990). Contingent valuation, experimental and theoretical research have all provided examples of situations where the first of these assumptions is simply not correct.[18]

Implications for Measurement

The valuation issues relevant to NRDA have direct implications for applying both indirect and direct methods to estimate these values. They affect what can be measured, how much experience we have had in practice with doing such measurement, and ultimately, the choice of an estimation strategy. Since indirect methods rely on reconstructing sufficient information from behavior to recover valuations, both the perspective (*ex ante* versus *ex post*) and the definition of what can be measured will depend on the analyst's assumptions about people's behavior and how it can be observed. For example, it is generally assumed, that travel cost demand models measure *ex post* use values.[19] When the hedonic property value model is used to estimate the marginal value of a resource's services, by comparison, the interpretation of the valuation concept will depend upon how the measure of services is defined. McConnell (1990), for example, suggests that when releases of hazardous waste are assumed to increase the risk facing an individual, this marginal value is an *ex ante* value.[20]

Neither of these indirect methods can estimate nonuse values. Only the direct survey approach (or CVM) can be used to measure nonuse values.

Interpreting the valuation estimates from some variant of contingent valuation will depend upon how the questions are posed to and interpreted by respondents. Nonetheless, it is probably best to treat the responses as *ex ante* values since CVM generally asks for anticipated values or behavioral responses. The distinction between whether the survey will include use and nonuse components will vary with the framing of the questions posed.

What we can expect from each method in practice will vary with the problem. Consider first the indirect methods--the travel cost recreation demand and hedonic property value models.[21] Application of either method to damage measurement focuses the analysis on what is generally considered to be the primary research issues for each method. For the travel cost model, the focus is on valuing changes in site quality. There are two important features of this task: (1) measuring quality in a way that parallels people's perceptions and can be connected to the natural resource injury; and (2) observing behavioral responses to quality differences that are comparable to people's responses to the release.

Applications of the hedonic property model face similar problems. However, these applications usually relate to housing prices in areas affected by a resource before and after a release. The key issue in the use of the hedonic method involves defining measures of the effects of a release that can be recognized by the in the market participants and that provide a reasonable index of how the resource's services would be affected.

Using any indirect method for measuring what an individual would pay to alter the effects of a release will be influenced by the analyst's judgment in defining how the release caused some change in variables that can be included in the models that are estimated from available data. These variables must provide an authentic description of the perceived impacts of injuries in order for the damage estimates to be acceptable.

The challenges facing contingent valuation or the survey approach are equally imposing. The practical issues involved will depend on the specifics of the case. Releases extending over long time periods, such as the case of the Eagle River, can easily involve baseline conditions outside the range of people's experience. Thus, a problem comparable to the measurement question for indirect methods (observing a comparable quality change) arises in explaining the "commodity change" to potential respondents for direct methods. When the release is recent, the valuation task requires separating respondents' perceptions of the appropriateness of the source of the change (see Fischhoff and Furby, 1988) from their valuation of what has changed. It is important that respondents separate the source of the damage from the effect of the damage on the natural asset, therefore, avoiding any bias in the value statement due to the "source of the change."

An Example: The Eagle River

The judgments involved in developing and using a model for natural resource damage assessment can take many forms, therefore, an actual case involving a mining site will be discussed to illustrate the effects of these decisions. The case was initiated before the DOI Type B rules were announced and has subsequently been settled. It involved the Eagle Mine and a five-mile section of the Eagle River between Gilman and Minturn (see Figure 1). The trustee (the State of Colorado) for the natural resources involved (public lands adjoining the river, the Eagle River, and groundwater aquifers near this section of the river) contended that past operations of the mine had resulted in releases of hazardous substances into the river and the groundwater. While the mine was no longer operating, the disposal of mine tailings and the condition of the old mine allowed continued releases to take place, with alleged injuries to the natural resources.

Comparing how the plaintiff (the trustee) and the defendant's analysts estimated the natural resource damages from these releases illustrates the importance of analyst judgment in applying valuation methodologies. Neither side attempted to develop comprehensive estimates for each of the natural resources, though the defendant's analysis is easier to associate with the two primary resources with alleged injuries--the river and the groundwater.

The plaintiff's analyst conducted two household mail surveys, one for the residents of Eagle County and the other for the entire state. The county survey collected information to implement three methods for measuring components of the damages. It asked:

(a)   how many days each respondent would spend in fishing and nonwater-based recreation activities if the section of the river (identified with the map in Figure 1) were restored to its "pre-mine" condition;

(b)   how much each respondent would be willing to pay annually for each of ten years to clean up this section of the Eagle River; and

(c)   if respondents were homeowners, the purchase price for their homes, a brief set of the housing characteristics, and the date of purchase.

Responses to the first question with measures of current water and nonwater activities and the U. S. Forest Service unit day values for these activities were used to estimate per person annual use values. The second question provided estimates per person of *ex ante* use and presumably nonuse values as a composite. The information on housing prices was used only for those individuals within 25 miles of the river, and a hedonic price function (using deflated prices) was estimated. A qualitative variable indicating location within six miles of this section of the river was assumed to reflect the effects of the releases into the river.

The defendant's (Gulf-Western Industries) analyst developed a travel cost recreation demand model using the U. S. Fish and Wildlife's 1980 Survey of hunting and fishing decisions. A subsample of individuals from the survey using fishing sites within a five-county region around the mine site was used to estimate the model. While the specific estimates of demand functions were not presented in the report prepared for the defendant, sufficient information was reported to highlight several aspects of this strategy. Two important assumptions in the defendant's analysis follow from implicit assumptions in their sample and their characterization of how the release would affect potential fishing on this section of the river.

First, by combining all the fishermen using sites anywhere in the five-county area around the mine, they implicitly assume that the sites are perfect substitutes for each other since all are assumed to be in the same market. Second, the analysis maintains that damage to the river's ability to support fishing implies recreationists will go to the next best alternative--the river above the mine site. This would imply traveling a maximum of five miles further each way. Thus, the loss of the site is described as a price change. However, the key assumption is that the choke price equals the implied travel cost to what is treated as a "perfect substitute" site--the river above the affected region.

The plaintiff's analysis, by contrast, tended to highlight the uniqueness of the site by asking county residents about their increased use of it without asking where that "increased use" would come from and whether it represented "new" recreational trips. The valuation measures implied for site services (per unit) by each side's model were actually quite comparable. The unit day values (in 1985 dollars) for water and nonwater based activities were $14 and $9 per day respectively,while the consumer surplus from the travel cost demand models for each type of activity were actually higher at $21 and $32 per day.

Comparisons of the annual values from the WTP component of the survey were also quite similar--with the county survey's CVM estimates of $73 per year for water-based and $51 per year for nonwater-based activities. The defendant's model estimated 6.18 days of fishing per season and 10 days for nonwater-based activities. Applying each to the average of the annual consumer surplus from fishing and nonwater-based activities, the defendant's analysis would imply values that <u>exceed</u> the CVM estimates. <u>However, the characterization of what was lost from the release is what distinguished the analysis at the level of an individual recreationist.</u>

By treating the release as requiring that recreationists use the next best alternative and incur <u>only</u> a price increase equivalent to traveling ten miles further, the per individual loss from the defendant's travel cost model becomes $1.35 per day for fishing and $.55 per day for nonwater-based activities (in

1985 dollars).

What may be more surprising is that these estimates would still exceed the per person annual values estimated by the plaintiff from the second contingent valuation survey intended to be representative of the state population's valuation of the river's services. This second survey progressively focused the valuation tasks, first asking annual WTP for cleanup of all 200 potential problem sites (again for each of ten years), then the percentage to be assigned to seven sites (including the mine) specifically identified and described in the survey, and finally a percentage of that amount for the site identified as most important.

The average annual per household values from the plaintiff's statewide contingent valuation survey included use and nonuse values. This estimate ($5.60) is less than the annual use values implied for fishing by the defendant's analysis ($1.35 x 6.18 days = $8.34) and very close to those for nonwater-based activities ($.55 x 10 days = $5.50).

Nonetheless, these comparisons offer a misleading impression of the disparity in each side's estimates of the present value of future losses if restoration did not take place. The plaintiff's estimates range from $15 to $45 million depending on which method is used to estimate per person losses and the treatment of nonuse values. Even the lowest end of this range is about 84 times the size of the defendant's estimates of $139,500 for fishing and nonwater-based activities. Moreover, the differences in discount rates, time horizon, and real growth of these values do not explain these differences.

They arise from one strategic assumption--each analyst's assumption about the extent of the market for the Eagle River. The plaintiff's analysis assumed the market corresponds to every household in Colorado on the grounds that their statewide survey was intended to represent this group. The defendant's analysis implied that the recreationists assumed to be currently involved in fishing and nonwater uses of the area would experience a gain (in the form of the reduced price) from restoring this section of the river. The difference cannot be attributed to the aggregate of nonuse values because these are already removed by focusing on the lowest estimate.

Research Issues

NRDA seems likely to influence the direction of nonmarket valuation research. In some cases, it will reinforce the attention already given to specific aspects of valuation measurement, such as estimating the value of changes in the quality of a resource's services. However, the cases settled or in progress to date suggest that it has identified new issues.

Based upon the perspective of narrowing the differences in damage assessments, the most important of these issues appear to be two aspects of defining the extent of the market. First, we need to develop better understanding of the implications of delineating the exact service to be valued. Does the injured resource provide services that have exceptionally close substitutes already available? If so, then a modeling strategy that focuses on valuing changes in access to these services would appropriately characterize how a resource injury creates economic losses.

Second, we need to identify the equivalent of a geographic extent of the market. This means the analysis must uncover which people have the measured values and which do not. It is no accident that these questions underlie the substantial differences between plaintiff and defendant's estimates of damages for the Eagle River case. Moreover, their relevance is not limited to that one case. Indeed, in a quite different context, the same questions play a central role in antitrust evaluations of whether changes in the number of firms or of products available in a market allow any particular firm to "price profitably."

Finally, consideration of both aspects of defining the extent of the market are relevant to both use and nonuse values. Indeed, as Figure 1 highlights, it may not be possible to use strategies that isolate market definition for use values from the effects of injury to a natural resource on each individual's nonuse value.[22]

COMPLETE RESTORATION, STRICT LIABILITY, AND EFFICIENCY

As part of the court of appeals ruling on DOI's procedures for Type B assessments, some components were remanded to DOI for modification. A common theme among the comments prepared in response to the Advance Notice of Proposed Rulemaking (issued September 22, 1989) is that some important aspects of the court's ruling were incorrect. In particular, several economists commenting on the ruling argued that using the costs of restoration to measure natural resource damages was incorrect (see Polinsky and Shavell, 1989, and Cicchetti, 1989). Polinsky and Shavell conclude their recommendations by observing that:

> The strict liability rule leads in principle to optimal responses to oil spills, to optimal precautions to prevent spills, and to optimal business activity. In contrast, the lesser of rule and the complete restoration rule each lead to inappropriate spending in response to spills, to excessive precautions to prevent spills, to curtailment of desirable business activity (p. 18).

This argument is only correct within a specific objective function that characterizes what is optimal. As Polinsky and Shavell note, it requires that we assume "...society's goal is to maximize the net value of the natural resource, that is, the value of the resource minus the costs of cleanup, restoration, and replacement" (p. 4).

NRDA entails an explicit assignment of property rights to natural resources to public trustees. It is not a policy designed to induce efficient production and transportation of oil and hazardous materials. It is certainly possible to abstract from the legislative mandates and ask a more general economic question. How do liability rules perform in inducing efficient resource allocations when the activities give rise to stochastic externalities? While this is separate from NRDA, it is also an important question (see Shavell, 1984, and Segerson, 1987). In failing to make these distinctions, past commentaries have confused analysis on evaluating the appropriate compensation for a wealth effect (loss) experienced by the public with the efficient use of compensation for restoration after it has been collected from responsible parties.

Considering the efficient use of compensation, the Court's ruling mandates policies that may not efficiently use the compensation to provide natural assets to the public.   Thus, the Polensky-Shavell or Cicchetti recommendations to undertake restoration until marginal benefits equal marginal costs define an efficient restoration plan, provided the goal is to maximize the net benefit derived from applying the recovered damages. Because the value of what remains will in general be less than the public's holdings of natural assets before the release, additional compensation is warranted if maintaining the real value of society's natural assets is the objective. An upper limit to this added compensation should be the total value of future services that will be lost after restoration.

Finally, judgments on the efficiency of rules that require full compensation should be evaluated from two related perspectives. First, Polinsky and Shavell observed, the incentive effects will promote inefficient responses, provided we analyze them within a static framework that ignores the goal of maintaining the value of society's endowment of natural assets (as a constraint to production and consumption activities). Then, the extent this is incorporated as a constraint on allocation decisions, strict liability with damages defined following Polinsky and Shavell (or Cicchetti) will fail to provide sufficient incentives to maintain the sustainability constraint for natural assets.

NRDA AND THE MINERAL INDUSTRY

The extractive sector probably generates a larger volume of solid waste residuals than any other production activity. In most cases, these residuals contain materials that would be considered hazardous wastes if leaked into rivers or aquifers. The Colorado Attorney General's office has estimated that as many as 200 sites in the state may have released hazardous wastes that are likely to have injured one or more natural resources (and therefore would qualify for a NRDA). This same pattern characterizes many other western states.   For example, in western Montana, there are currently four separate but contiguous Superfund sites on the National Priorities List in the Clark Fork Basin alone. The Silver Bow Creek and Upper Clark Fork currently compose one of the largest and most complex Superfund sites in the nation.

Initially, the authority to act as trustee was limited to the federal government. This has been progressively expanded to include all levels of government and Indian tribes. Considering all the resources covered under natural resource damage liability, Breen (1989) has estimated that there are about 69,000 potential plaintiffs.

Thus, NRDAs are likely to play an important role in influencing how current mining firms handle their operations and dispose of mining residuals. Also, since the liability is joint and several, it implicitly allows current claims for damages from any past operations that led to releases. This was clearly the situation for Gulf-Western Industries, a firm no longer involved in the extractive sector but nonetheless held responsible for damages at the Eagle River.

The doctrine of joint and several liability also has important implications for acquiring property that may have been associated with mining activities. With purchase comes the potential for future liability. Thus, economic assessments of potential damages could well become part of the routine appraisals of properties involved in such activities and may play a role in determining sale prices. Therefore, estimates of nonmarket valuation could have a direct effect on market prices. With sufficient market activity, one could envision a market's disciplinary force imposed (through opportunities for arbitrage) on the practices used in nonmarket valuation.

One important academic implication of this new legal environment (in addition to the influences already mentioned on research regarding nonmarket valuation methods) is likely to be a reorientation in the methods used to describe the privately efficient behavior of extractive firms. Conventional practice has tended to focus most of its attention on how ore deposits as assets affect the long-term viability of a firm. Static resource allocation decisions were incapable of describing how this connection influenced the optimal extraction profile.

Modern versions of this model consider the role of "static" pollution (externalities with effects limited to individual periods). NRDA is likely to increase interest in treating the firm's problem as one involving the management of multiple natural assets--some providing extractive outputs and others yielding amenity services. While the latter cannot be sold on organized markets, natural resource damage liability and the need to dispose of residuals generated from the extractive activities serve to define a new stock constraint, and with it a new generation of research and policy issues.

Figure 1: Map of the Eagle Mine and Eagle River

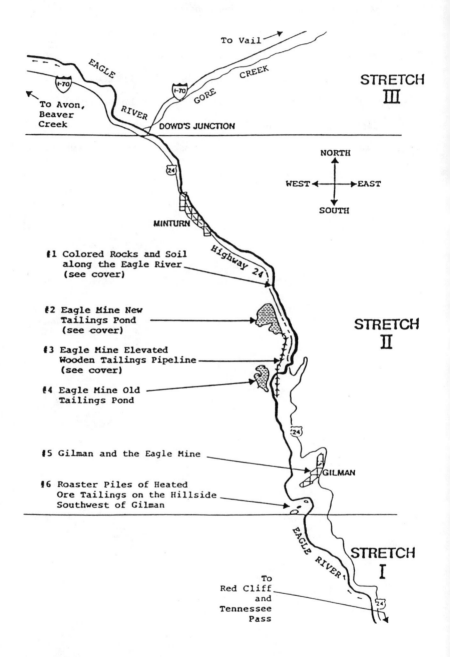

Source: Rowe and Schulze (1985)

NOTES

*University Distinguished Professor, Department of Economics, North Carolina State University Fellow, Resources for the Future. Thanks are due Ray Kopp for numerous insights in our joint research in this area, Greg Michaels for his assistance in understanding the RCRA amendments, John Tilton and seminar participants at the Colorado School of Mines for helpful comments on an earlier draft, and Barbara Scott for her careful editing of earlier drafts of this paper.

1.   There are several examples that oil firms believe these renewed concerns over security of foreign oil supplies will reduce the public's resolve to avoid coastal drilling and exploratory activity in the Arctic National Wildlife Refuge.

2.   The most recent round involves an EPA study, Report to Congress on Special Wastes from Mineral Processing, completed July 31, 1990, that has been challenged by environmental groups and the treatment industry.

3.   For a complete description of the evaluation of natural resource liability, see Breen [1989]. He argues that this doctrine has steadily expanded from its beginning in the 1970s to the 1988 amendments to the Marine Protection, Research and Sanctuaries Act. Expansion has occurred in four primary areas with increases in:

(a)   the number of total potential plaintiffs from the federal government to some 69,000 as of current legislation;
(b)   the context for suits to be brought;
(c)   the types of substances or damage no matter how inflicted (in the Marine Protection amendments); and
(d)   the components considered in the measure of damages.

4.   See Kopp and Smith, forthcoming RFF, for a more complete discussion of the legislative and administrative history of DOI actions.

5.   It is not clear whether a plaintiff can have access to rebuttable presumption, the primary gain from using the rules, until a final set of rules has been proposed by DOI.

6.   Natural resources are defined (in both the legislation and in the DOI rules following the CERCLA/SATA legislation) in very broad terms to include"...land, fish, wildlife, biota, air, water, ground water, drinking water supplies and other such resources belonging to, managed by, held in trust by, appertaining to, or otherwise controlled by the United States (including the resources of the fishery conservation some established by the Magnuson Fisher Conservation and Management Act of 1976)..." (Federal Register, Volume 51,

No. 148, August 1, 1986, Rules and Regulations, Part III, Department of the
Interior, p. 27727).

7.   See Brown [1990] for a discussion of the problems caused by economic
damages arising from the perception of injury that could not be claimed without
establishing scientifically an injury to one or more components of the natural
resource.

8.   The Court of Appeals ruling specifically notes that:

>   The statute's command is expressly not limited to use value;
>   if anything, the language implies that DOI is to include in its
>   regulations other factors in addition to use value...option and
>   existence values may represent 'passive' use, but they
>   nonetheless reflect utility derived by humans from a resource,
>   and thus, *prima facie* ought to be included in a damage
>   assessment" (p. 67).

9.   Recently Larson [1990] has demonstrated and extended an argument by
Neill [1988] that if a nontraded good can be related to a set of market goods and
preferences are implicitly separable, the marginal willingness to pay for changes
in that good can be recovered. Because weak complementarity is not required,
this value will include (*ex post*) use and nonuse values. However, as Larson
clearly acknowledges, this result will be sensitive to the specification of how the
nontraded goods affect the demand system for the traded goods.
        See also Smith [1991] for a discussion of how Neil's work relates to
household production models and applications that can be interpreted as
implementing Larson's suggestions.

10.   The Court of Appeals explained its rejection of the DOI "lesser of" rule
in detail. The key discussion highlights this point, noting that:

>   The fatal flaw of Interior's approach, however, is that it
>   assumes that natural resources are fungible goods, just like
>   any other, and that the value to society generated by a
>   particular resource can be accurately measured in every case--
>   assumptions that Congress apparently rejected. As the
>   foregoing examination of CERCLA's text, structure and
>   legislative history illustrates, Congress saw restoration as the
>   presumptively correct remedy for injury to natural resources.
>   To say that Congress placed a thumb on the scales in favor of
>   restoration is not to say that it forswore the goal of efficiency.
>   'Efficiency' standing alone simply means that the chosen
>   policy will dictate the result that achieves the greatest value to
>   society. Whether a particular choice is efficient depends on

how the various alternatives are valued.   Our reading of CERCLA does not attribute to Congress an irrational dislike of 'efficiency';   rather it suggests that Congress was skeptical of the ability of human beings to measure the true 'value' of a natural resource" (pp. 50-51).

11.   The ruling does not indicate what would constitute grossly disproportionate costs.   It does indicate as an example that DOI has the authorization to specify this as some multiple of the economic value of damages. Among several examples of how this definition might be given, the ruling indicates it could be stated as three times use value (p. 22).

12.   The identified methods were market prices for the resources as assets or appraisals of these prices.

13.   For a more complete discussion of both the direct and indirect methods for valuing nonmarket goods see Appendix A.

14.   Quiggin [1990] makes this point but argues it is the reason why we cannot incorporate existence values into conventional applied welfare economics.

15.   See Smith [1990] for a general discussion of the importance of a behavioral trail in benefit measurement.

16.   A more technical discussion is provided in Appendix B.

17.   In the ex ante situation, the release may also affect the probabilities associated with the expected utility and a use value could be defined for policies that avoid the change in these probabilities.

18.   Hanemann's [1990] recent analysis suggests that the relative size will depend on the existence of substitutes for the resource as well as the size of income effects.   For reasonable assumptions on income elasticities and elasticities of substitution with plausible utility functions, the ratio of WTA/WTP can be as large as five.

19.   It could be possible to identify a travel cost model with an ex ante value if the model was based on data requesting respondents' expected or anticipated trips during a time period or if the source of the uncertainty was one or more site characteristics that would not be known before traveling to the site (i.e., congestion, quality of water, weather conditions, etc.).

20.   This relationship between the incremental option price for a risk change and the slop of a hedonic model with that risk as an argument was initially proposed in Smith [1985].

21.   See Appendix A for a brief review of the logic of each method and the
issues to be considered in implementing each.

22.   See Bergstrom, Blue and Varion [1986] for a discussion of the
connections between decisions at the extensive and intensive margins of choice
in evaluating mixed private and public provision of pure public goods.

REFERENCES

Bergstrom, Theodore, Lawrence Blum and Hal Varian. 1986. "On the Private Provision of Public Goods." Journal of Public Economics 29 (No. 1, February): 25-50.

Breen, Barry. 1989. "Citizen Suits for Natural Resource Damages: Closing a Gap in Federal Environmental Law." Wake Forest Law Review 24: 851-80.

Brown, Gardner M. 1990. "Economics of Natural Resource Damage Assessment: A Critique," in Valuing Natural Assets: The Economics of Natural Resource Damage Assessment, R. J. Kopp and V. K. Smith, eds. Unpublished edited manuscript under revision for publication by Resources for the Future.

Cicchetti, Charles J. 1989. "Comments on the U.S. Department of Interior's Advanced Notice of Proposed Rulemaking 43CFR Part 11 Natural Resource Damage Assessments. Putnam, Hayes and Bartlett, Inc. (November 13).

Cummings, Ronald G., David S. Brookshire, and William D. Schulze. 1986. Valuing Environmental Goods: An Assessment of the Contingent Valuation Method (Totowa, N.J.: Roman and Allanheld).

Fischhoff, Baruch and Lita Furby. 1988. "Measuring Values: A Conceptual Framework for Interpreting Transactions With Special Reference to Contingent Valuation of Visibility." Journal of Risk and Uncertainty 1 (June): 147-184.

Hanemann, W. Michael. 1990. "Willingness to Pay and Willingness to Accept: How Much Can They Differ?" American Economic Review (in press).

Jones, Carol Adaire and Yuc Sheng Sung. 1990. "Use of Discrete Choice Models to Value Natural Resource Damages in Recreational Fisheries." Paper presented at American Agricultural Economics Association Meetings, Vancouver, British Columbia (August 4-8).

Kopp, Raymond J. and V. Kerry Smith. 1989. "Benefit Estimation Goes to Court: The Case of Natural Resource Damage Assessments." Journal of Policy Analysis and Management 8 (Fall): 593-612.

Kopp, Raymond J., Paul R. Portney, and V. Kerry Smith. 1990. "The Economics of Natural Resource Damages After Ohio v. U. S. Department of the Interior." Environmental Law Reporter 20 (No. 4): 10127-10131.

Larson, Douglas M. 1990. "Measuring Willingness to Pay for Nonmarket Goods." Paper presented at American Agricultural Economics Association Meetings, Vancouver, British Columbia (August 4-8).

McConnell, K. E. 1990. "Indirect Methods for Assessing Natural Resource Damages Under CERCLA," in Valuing Natural Assets: The Economics of Natural Resource Damage Assessment, R. J. Kopp and V. K. Smith, eds. Unpublished edited manuscript under revision for publication by Resources for the Future.

National Economic Research Associates. 1985. State of Colorado et al. vs. Gulf & Western Industries, Inc. et al.: Response to ERC's Report on Damages. White Plains, N.J. (December).

Neill, Jon R. 1988. "Another Theorem on Using Market Demands to Determine Willingness to Pay for Non-Traded Goods," Journal of Environmental Economics and Management 15 (June): 224-232.

Polinsky, A. Mitchell and Steven Shavell. 1989. "Economic Analysis of Liability for Natural Resource Damages Caused by an Oil Spill." Comments submitted to U.S. Department of Interior in response to the Advanced Notice of Proposed Rulemaking for Type B Natural Resource Damage Assessments (November 10).

Quiggin, John. 1990. "Do Existence Values Exist?" Unpublished paper, Department of Agricultural & Resource Economics, University of Maryland, (March).

Rowe, Robert D. and William D. Schulze. 1985. Economic Assessment of Damage Related to the Eagle Mine Facility. Energy and Resource Consultants, Inc., Boulder Colorado (November 27).

Segerson, Kathleen. 1987. "Risk Sharing and Liability in the Control of Stochastic Externalities." Marine Resource Economics 4 (3): 175-192.

Shavell, Steven. 1984. "Liability for Harm Versus Regulation of Safety." Journal of Legal Studies 8 (June): 357-374.

Smith, V. Kerry. 1985. "Supply Uncertainty, Option Price and Indirect Benefit Estimation," Land Economics 61 (August): 303-308.

Smith, V. Kerry. 1987. "Nonuse Values in Benefit Cost Analysis." Southern Economic Journal 54 (July): 19-26.

Smith, V. Kerry. 1990. "Can We Measure the Value of Environmental Amenities?" Southern Economic Journal 56 (April): 865-878.

Smith, V. Kerry. 1991. "Household Production Functions and Environmental Benefit Estimation" in Environmental Benefit Measurement, eds. John Braden and Charles Kolstad (Amsterdam: North Holland, in press).

U. S. Court of Appeals, District of Columbia. 1989. Ohio vs. United States Department of the Interior 880 F.2d 432, 19 ELR 21099 (D. C. Cir, July 14, 1989).

U. S. Department of Interior, Office of Secretary, 43CRF Part 11. 1986. "Natural Resource Damage Assessments: Final Rule," Federal Register 51 (No. 1148, August 1), 27673-27753.

U. S. District Court, District of Massachusetts. 1989. "Memorandum Concerning Natural Resource Damages Under CERCLA concerning Acushnet River and New Bedford Harbor Proceeding Re Alleged PCB Pollution," by Judge D. J. Young (June 7).

APPENDIX A: NONMARKET VALUATION METHODS

This appendix describes four methods used to estimate the monetary values people realize from nonmarketed goods or services. Three of these approaches are indirect or observed behavior methods and the fourth is the direct or survey method. The first class includes travel cost demand, hedonic, factor income, and averting behavior models. In each case, these methods use information on the actions of individuals (or firms), along with assumptions about what motivates those actions, to estimate an individual's implied marginal value for an improvement in the resource. The specific assumptions used to recover these estimates vary with the modeling framework used, as well as the information available. The two most commonly used approaches in this class for natural resource damage assessments are the travel cost demand and hedonic property value models.

The first is the conventional approach for estimating the demand for recreation sites. Beginning with Harold Hotelling [1947, Letter to National Park Service in An Economic Study of the Monetary Evaluation of Recreation in the National Parks, U. S. Department of the Interior, National Park Service and Recreational Planning Division, 1949], this framework relies on a simple insight. Visitors pay an implicit price for the use of a recreation site in the form of the travel and time costs associated with gaining access to the site. Thus these costs, together with any entrance fees, serve (for a fixed length and single objective trip) as an implicit price for a site's services. By observing site usage from different distances, this method has proved to be exceptionally robust for estimating the demand for recreation sites of all types. Early applications involved data in aggregate form based on origin zones. More recently, on-site surveys of users have provided micro data on individuals' patterns of use of specific sites and their costs. However, these new data sets have created a new range of econometric issues by virtue of failing to observe individuals who decide not to visit each specific site. The actual modeling has largely been single equation demand models. However, some authors have used these types of data to estimate random utility models, treating each visit as a decision made independently from all previous trips.

The second popular indirect method for estimating the value of nonmarketed resources is the hedonic model. This framework relies on two key assumptions to estimate the marginal value of an increase in a positive environmental good or a decrease in a negative environmental externality. The first assumption involves some clearly recognized (to market participants) technical association between the nonmarketed commodity (or a reliable proxy variable assumed to represent the commodity) and the property whose prices are being analyzed. The second assumes that the property market "linked" to the commodity is sufficiently open to assure that housing trades will continue until prices provide no incentive for change. Because the commodity involved (housing) is very heterogeneous, the model predicts that a set of prices will be required to generate an equilibrium matching of buyers and sellers. This set is usually assumed to be large enough to be approximated by a continuous function relating the equilibrium prices to the characteristics of each house.   Jan Tinbergen [1956, "On the Theory of Income Distribution," Weltwertschaftliches Archiv, Vol. 77, pp. 155-175] was one of the first to analytically derive an expression for this price function in the context of a labor market equilibrium.

This equilibrium assures (for individuals who can choose any set of small changes in the housing attributes important to them at the time they select a house) that the derivative of the price function with respect to each nonmarketed commodity will provide estimates of the marginal value of that commodity (expressed as the present value).   Under certain circumstances, these marginal values can be used to estimate the full inverse demand function for this nonmarketed good.

The remaining indirect approaches, the factor income and averting behavior methods, use assumed connections between the nonmarketed commodity and production or cost relationships to estimate the value economic agents derive from the changes in the terms of access, quantity, or quality of the commodity involved.

The direct approaches involve just that--direct questions of individuals about how they would value some change in the terms of access or quality of a resource.   This process involves surveys of households using either personal interviews, telephone surveys, or mailed questionnaires.   The last two approaches are the most frequent in current use (because of cost considerations).

After over two decades of experience, this method has gained more widespread acceptance among conventional economists.   The valuation questions have been asked in a wide array of ways--bidding sequences, direct one-response values, yes/no or closed-ended surveys, rankings of combinations of commodities and payments, as well as several other newer forms.   These approaches require less theoretical assumptions to recover valuation estimates, but do imply that responses to hypothetical questions will, if properly framed, authentically characterize actual values.

Unit day values, as they have evolved, draw from each of these approaches.   They generally involve groups of experts attempting to interpret from the existing set of estimates (regardless of method used in the original study) a best estimate for each of a set of generic types of environmental resources or activities.   Therefore, this approach can combine findings from each of the above and relies on converting these to a standardized format per day of each type of activity in an attempt to provide an approximate value for the resources supporting these activities.

Table 1A summarizes some of the key issues involved in using these models and what is known about the "best" practices to resolve them.

Table 1A: A Checklist for Evaluating Indirect and Direct Approaches for Benefit Measurement

Method	Implementation Issue	Effects/Implications

I. Indirect Approaches

Method	Implementation Issue	Effects/Implications
A. Travel Cost Recreation Demand	Type of Data: Origin zone aggregate versus individual interviews	Origin zone does not allow measurement of time costs of travel or features of visitors; all determinants subject to significant measurement errors with these data; spatial dimensions important. Individual on-site interviews raise estimation problems--truncation and selection effects.
	Specification of Function for Demand	Estimates of value of recreation site's services will be sensitive to form selected for demand function.
	Key Determinants to Consider	Most important influences to benefit measures appear to be definition of dependent variable (trips versus days); treatment of the cost of travel time; description of role of substitute sites). Omission of these variables will bias demand estimates and can bias benefit estimates.
B. Hedonic Property Value Model	Type of Data: Census tract reports of owner's estimates of price versus sales	Sales price is always preferred. Census tract information not consistent with assumption that prices reflect market. Preferable for sample to avoid screening by mortgage source or other criteria.
	Measurement of Environmental Amenity	As a rule technical measures are available for amenities; what influences choice is perception. Connection between what is measured and what can be perceived is crucial.
	Description of Housing Neighborhood: Local finance/public goods characteristics	Model can be very sensitive to specification of determinants.
	Specification of Hedonic Function	Model assumes prices are result of an equilibrium. Function describes price/characteristic relationship required for equilibrium. Simulation results favor searching flexible forms.
	Participant Knowledge	Essential to document participants ability to observe (or know about) location-specific differences in amenities or characteristics in model.

Table 1A continued.

Method	Implementation Issue	Effects/Implications
	Extent of Market	Housing market may be composed of submarkets; need to investigate factors "segregating" potential demanders and suppliers.
	Benefit Measurement Confined to Marginal	Hedonic models measure at best one value for individual's marginal value of amenity. To recover full schedule requires additional information from theory or multiple hedonic models.
C. Averting Cost	Definition of Activity	The motives for activity must be exclusively associated with obtaining more (or avoiding reductions) in amenity to be valued; otherwise cost allocation problems.
	Specification of Function	Use of production/cost/discrete choice models depends on what is available. Process inherently <u>ad hoc</u> because of limited experience with models; failure to recognize prospects for economic adjustment is a serious source of error.
II. Direct Approaches		
A. Contingent Valuation Surveys	Framing of Questions	Description of hypothetical scenario is crucial to plausibility of responses; depends on mode of interview (mail, phone, in-person).
	Survey Design	Sampling practice; assurance of high response rates; explicit definition of target population.
	Analysis Protocol	Requires documentation of no response to specific questions; cross-checks for rejection or misunderstanding; tests and calibration with actual market-based decisions; testing of behavioral framing, interview effects.
B. Contingent Behavior (same issues as Contingent Valuation)	Experimental Design	Design range of values of key factors to vary across respondents.
	Behavioral Model	Responses must be analyzed in context of specific, behaviorally consistent model of respondents' decisions in relation to questions asked.

APPENDIX B: CLASSIFYING NATURAL RESOURCES VALUE

Figure 1B illustrates how choosing a treatment for uncertainty affects the interpretation of valuation measures and the separation of these measures into use and nonuse components. *Ex post* valuation measures relate to situations where the analyst assumes little uncertainty exists for the typical individual's decisions. Thus, uncertainty does not have important behavioral significance.[1] The analytical shorthand developed in the past literature to describe an economic model's distinctions between use and nonuse values requires that the model identify private commodities as X_i (with i designating the ith individual) and that it distinguish three aspects of the services of a natural resource--an individualized consumption, designated as r_i; a component that is analogous to a public good, R; and a quality feature, q. To measure the damages associated with an injury to a natural resource, that injury must be hypothesized to influence one or more aspects of these services.

Monetary measures of the value of some change in the parameters to an individual's decisions can only be defined within the context of a model of these decisions. Thus, Figure 1B indicates that each person is assumed to maximize **utility** subject to constraints in definite situations; while in uncertain ones, the conventional objective function calls for maximizing **expected utility**. The representation given in the figure uses Π_j to identify the probabilities of states of nature ($\overline{\Pi}$ as a vector for all probabilities) and assumes preferences will be state dependent. (This is designated with the subscript j.) The influence of constraints on behavior is indicated through the indirect utility functions for each situation. They have different symbols (V and v) because each corresponds to a different conception of individual well-being (utility in cases where uncertainty has been assumed to have minor behavioral significance and expected utility when it does). The arguments in each include for simplicity the same prices (P_x and P_r) and income. These parameters also could be assumed to differ between the definite and uncertain situations.[2] While the analytical description does not indicate it, most applications of either formulation of the decision process assume people will have different preferences.

Figure 1B: Conceptual Framework for Defining *Ex Ante* and *Ex Post* Use and Nonuse Values

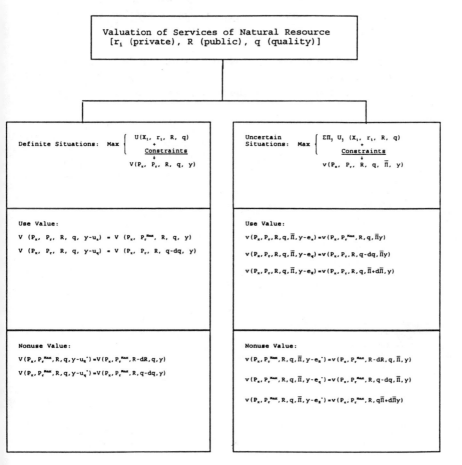

NOTES

1.   The central issue in defining behavioral significance concerns whether individuals could take actions to respond to uncertainty that would affect its impact on them.

2.   If some form of contingent claims markets can be assumed to exist, then the prices in the two specifications would be different. The presence of such markets offers one means of adjusting to the uncertainty.

CHAPTER
III

VALUING ENVIRONMENTAL DAMAGES WITH THE
CONTINGENT VALUATION METHOD: A CRITIQUE[1]

Ronald Cummings,
Philip Ganderton, &
Thomas McGuckin

OVERVIEW OF THE PROBLEM

Issues associated with natural resources damages, and how such damages might be measured as necessary for the Comprehensive Environmental Response, Compensation and Liability Act of 1980 (CERCLA), are of obvious contemporary concern to the mining industry and to resource/environmental economists. Over the past few years, the bulk of this concern has focused on the issue of how such damages might be measured, particularly damages relevant for Type B assessments which involve nonuse or, more generally, non-market elements.[2] For a broad class of these damages, there would appear to be but one method available for estimating their values: the Contingent Valuation Method (CVM).

The substance of values derived with the CVM has been a source of debate over the last ten to fifteen years (see Cummings, Brookshire and Schulze, 1986). The relevance of this debate has taken on new meanings over the last two years during which time the CVM has been elevated from a technique for valuing public goods with which economists were experimenting to "best available technology" for estimating *compensatory damages* in CERCLA adjudications. Understandably, the consonance between values derived with the CVM and market-like values which reflect a real economic commitment on the part of individuals has become an issue of central importance in contemporary debates concerning the substance of the CVM.[3]

The purpose of this paper is to comment on three aspects of the CVM which we regard as being central to contemporary considerations of the substance of CVM values. Our comments will be relatively brief and references to the existing literature will not be exhaustive. We will set out our views of the state of the art of the CVM as it relates to this issue. In presenting these comments, our purpose is to set out our basic position in the CVM debate, viz., that given the present state of the art of the CVM there exists little basis for the presumption that CVM values are in any way consonant with values actually held by subjects in CVM experiments for environmental "goods." This conclusion follows, we will argue, from our review of the literature related to three major questions, two of which involve behavioral issues and a third which involves a more technical, statistical issue. Section II will address the question: do CVM subjects understand the good which they are to value? The next section addresses the question: do CVM subjects understand the hypothetical market? Finally, section IV focuses on the question: is there substance in the empirical content of CVM values? Concluding remarks are offered in section V.

DO CVM SUBJECTS UNDERSTAND THE GOOD WHICH THEY ARE TO VALUE?

One finds in the CVM literature repeated references to the "strong" theoretical foundation underlying the CVM as a method for estimating values for non-market goods. In this regard, consider the following two examples.

> "Constructed markets (the CVM) enjoy a very strong theoretical foundation (...) Constructed markets, in principle and in contrast to other benefit measurement techniques, can

directly obtain willingness-to-pay or willingness-to-accept."
Carson [1991, p. 123]

"The reported empirical case (involving use of the CVM)
corresponds to the structure of the Hicksian equivalent
measure of willingness to pay." Hoehn and Loomis [1991, p.
2]

The theoretical foundation to which reference is made here is the economist's
received theory of value. This must surely be the case inasmuch as one does
not find in the literature a theory of consumer behavior *in hypothetical markets*
that is distinct from received theory.[4]

Is the theory of consumer behavior under conditions where both
transactions and payments are real applicable to behavior under the "market"
conditions of the CVM where both transactions and payments are hypothetical?
In the absence of any theory of hypothetical markets, this is surely an empirical
question. Therefore, it is necessary to inquire as to the extent to which there
exists empirical evidence which would support this notion.

In this regard, particular concern must be given to the assumption of
received theory which directly relates to the manner in which individuals are
assumed to form values for goods and services in conjunction with their
preferences for such goods and services. Received value theory assumes
(among other things) that individuals have "well-behaved" preferences over a
choice set. Specifically, it is assumed that preferences are *Complete* (any two
consumption bundles can be compared[5] in terms of utility), *Reflexive* (any
consumption bundle is at least as good as itself)[6], *Transitive* (if bundle A is
preferred to bundle B, and B is preferred to bundle C, then A is always
preferred to C), *Monotonic* (more of a good[7] is better than less), and *Quasi-
concave*[8] (implying that average consumption bundles are weakly preferred to
extreme bundles). In the present context we would add a requirement that the
consumer know the existing quantities and qualities of goods that he does not
have a choice over, since these can well affect his preferences.

It is necessary to carefully consider the implications of this assumption.

It implies that individuals have a firm understanding of goods which they are to value. Of all of the hundreds if not thousands of goods which an individual might acquire, some subset of these goods form the individual's choice set. The choice set consists of the goods which the individual is to acquire. The choice set may include relatively few or many goods; the individual's tastes, preferences and income determines which goods are included or excluded from this set (note the potential importance of information in this regard).

Obviously, the composition of choice sets may vary widely across different individuals. There may be many goods that provide utility to the individual, but that are excluded from his choice set. Examples include the many environmental and public goods from which the individual may derive satisfaction. He "knows," or understands, that these goods exist, he may be aware of the satisfaction that he derives from them, but they are not in his choice set for the simple reason that their quantities and qualities are fixed--they are not acquired in the way that other goods are acquired. Since he has no direct choice over the levels at which these public goods are provided, they are not a part of his choice set. For the subset of goods in the individual's choice set, his preferences across these goods are said to be complete, reflective, transitive, monotonic and quasi-concave.

It is also important to realize that the composition of an individual's choice set may change from time to time. Such changes may be frequent or infrequent. Goods once in the set may be discarded and new ones added. The central issue, however, is that an individual's determination of the composition of his choice set at any point in time is based upon his knowledge of goods and services which can be acquired. Goods which cannot be acquired are ignored by him as irrelevant. For example, an individual may have knowledge of an expensive car but not have the means (income) to acquire it, therefore, it is not in this individuals choice set. Given any change in the composition of his choice set, the individual must then re-evaluate all goods in the new consumption set to the end of determining his complete, reflexive, transitive, monotonic and quasi-concave preferences among the various goods. This re-evaluation process is very important. It involves the individual's assessment of, as examples, how various goods might be preferred over others, the extent to which one good or set of goods might be substitutable for other goods ("if I buy this good, I need not buy that good: they both basically serve the same purpose"), and the complementarity of goods ("if I buy more of this good, I may need to buy more of that good"). The implications of this re-evaluation process, referred to by some as the process of "researching one's preferences," will become a topic of interest below.

Summarizing the salient aspects of this key assumption from value theory highlight several points of interest. In value theory a consumers choice among goods is based upon the assumption that individual's are *aware of goods which they might acquire*, that their tastes and preferences guide them in their choice of goods to be included in a choice set, and that they have established "well-behaved" preferences for goods in this set. This setting for individual choice is known as a "well-defined choice set."

The use of CVM invites the subject to introduce a "new" good into his choice set. It may then be asked: is the individual's valuation for the newly introduced good derived from a well-defined choice set?

The importance of well-defined choice sets for true valuations, valuations which reflect a real economic commitment, is well recognized. The inextricable relationship between well-defined choice sets and true values is described by Freeman (1986) in the following manner.

> "It is conventional to assume that individuals have well defined preference orderings and that they know the shape of their indifference curves. Thus, if we observe an individual to accept a trade-off between income and some other good, we believe that he has revealed something about his preference ordering (...)the inference that revealed trade-offs reflect true valuings or preferences is correct *only if individuals do in fact have full knowledge of their preference orderings* (...) Suppose that due to (...) the introduction of a new good, an individual has an opportunity to choose from among a set of consumption

> bundles that are unfamiliar to her (...) It seems plausible that she might experiment with several different consumption bundles before settling into a new equilibrium position. This experimentation can be viewed as an effort to explore an unfamiliar part of her preference ordering. We can only accept revealed preferences as reflecting true preferences after this exploration has been completed. Therefore I want to define the true value of the environmental good as that substitution between income and the environmental good which we would observe after repeated trials or opportunities for the individual to alter her consumption position" Freeman [1986, p. 150].

The introspective process of preference research required for the individual to form a well defined choice set is then seen to involve two interrelated concepts: time and information. Particularly in instances where the new good is unfamiliar to the individual, time is required if the individual is to "explore," *a la* Freeman, unfamiliar parts of her preference orderings. The importance of time and information is also emphasized by Hoehn and Randall (1987) as follows.

> "The time and resource constraints of the CVM context may introduce two sources of error into the value formulation process. First, information errors may arise as complex policy information is communicated to the respondent by the CVM format. Errors may be left uncorrected due to time constraints on repetition and review. Thus, the time constrained process of communicating complete information may introduce an additional source of uncertainty into the policy scenario as perceived by the respondent. Second, once policy information is assimilated, the process of evaluation--of selecting a bid--may also be cut short by limited time and decision resources" Hoehn and Randall [1987, p. 229-230].

The time and preference research dimensions of an individual's well defined choice set are described by Eberle and Hayden (1991) in a different and interesting way: "How the market is presented to the subject may determine his beliefs about the market. This is especially a problem when the subject is required to value natural resources or an ecosystem the respondent does not understand and has usually never observed. Having no foundation in his memory, *the presented market becomes the basis for his decision* (emphasis added)" (ibid., p. 661-662).

The information requirement for a well defined choice set has another dimension of importance as well: information as to relevant substitutes and complements for the good being valued. Referring to the customary process by which benefit-cost studies for individual public goods are typically conducted in isolation (where substitutes and complements are not considered), Hoehn and Randall (1989) conclude that valuations which exclude relevant substitutes and compliments may be misleading.

> "For instance, consider only the problem of species preservation, a relatively small subset of environmental concerns. Conventional benefit cost logic demonstrates the nontrivial benefits for each of a limited number of representative species. There seems little reason to doubt that a similar level of benefits could be demonstrated for many other endangered species--at least when each is evaluated independently. Yet biologists suggest that there are literally hundreds of thousands of species in danger of extinction. Surely, in this sort of policy environment, conventional procedures overlook some crucial element of the evaluation problem." Hoehn and Randall [1989, p. 550]

Thus, to continue Hoehn and Randall's example, a well-defined choice set for the individual (whose value is at issue in the benefit-cost study) who is to add "preserve species X" to his choice set may require consideration of many other species whose preservation might substitute (given the individual's preferences) for species X. The implications of values derived from choice sets that are not well defined are straightforward. Hoehn and Randall (1989, p. 550) state that their analyses ". . . demonstrate that conventional procedures (which ignore substitutes) systematically overstate net benefits. . . ." Freeman rejects the presumption of "trueness" for values so derived. Certainly, valuation behavior under such circumstances would be clearly inconsistent with basic theoretical postulates which are claimed to underlie the CVM.

Do subjects in the CVM "understand" goods which they are to value as "understand" refers to a well defined choice set underlying reported values?

To answer this question, the relevant literature must be examined.

The first study of interest in this regard is the work of Hoehn and Randall (1989). In this study the authors point to biases in CVM value measures which may result in instances where subjects are asked to value an isolated effect associated with a posited environmental policy. Most environmental policies, they argue, are multidimensional in effect: they give rise to effects which may be viewed as substitutes or complements to the specific effect being valued with the CVM.

Hoehn and Loomis (1991) attempt to quantify such biases. Using a mail survey, they ask subjects to value ten related "goods": five specific environmental goods and 5 "packages" of environmental goods. Each "package" contains 3 of the 5 environmental goods. Their analyses demonstrate dramatic differences between valuations for a specific commodity valued in isolation and valuations for the commodity when it is placed within the context of compliments and substitutes: the former value may overstate subjects valuations by as much as 70% (ibid., 1991, Table 1).

A similar result is reported by Hoehn (1991). One group of Chicago subjects was asked to value an 83% air quality improvement in the Grand Canyon National Park. A second group of subjects was asked two questions. First, their valuation of a 100% air quality improvement in Chicago. Following their response to this question, they were asked to value a joint program: a 100% air quality improvement in Chicago and an 83% air quality improvement in the Grand Canyon National Park. The mean value for the program to improve air quality in the Grand Canyon National Park and in Chicago was $83.00/household and $179.00/household, respectively, when the programs were valued in isolation. The joint program was valued at $190.00/household, however, implying a value for air quality in the Grand Canyon of $11.00 in contrast to the $83.00 value obtained when it was valued in isolation (ibid., 1991, p. 296). These data, along with those reported in the Hoehn and Loomis (1991) study, suggest that CVM subjects are not aware of relevant substitutes in their framing of CVM values: the introduction of substitute goods results in substantively lower CVM values.[9]

Next the literature which focuses on a question that derives from the notion that CVM subjects base valuation reports on well defined choice sets is combined: do changes in the quantity or quality of information given subjects affect their CVM valuations? The notion that changes in information given CVM subjects will affect values reported for a particular environmental good is argued by Cummings, Brookshire and Schulze (1986, p. 54-55) and derived as a testable hypothesis from an extended model of consumer choice by Hoehn and Randall (1987). There are a number of studies which develop empirical results which may be taken as relating to the hypothesis that changes in information result in significant changes in subject's valuations.

Smith and Desvousges (1987) find significant differences in subject's marginal valuations of risk with declining risk levels that result from changing (information relating to) baseline risks of premature death. Indeed, in contrast to Jones-Lee, Hammerton and Philips's (1985) findings of declining marginal valuations of risk with reduced risk levels, Smith and Desvousges (1987) find increasing marginal valuations when subjects value risk reductions from varying baseline risk levels (ibid., 1987, p. 108).

Bergstrom, Stoll and Randall (1989) test the hypothesis that CVM values will increase in response to subject's being given "perspective" and "relative" information concerning the relationship between bids and incomes. An initial WTP is obtained from student subjects for facilities that provide access to a well known river in central Texas used for recreation. Subjects are then told the percent that this bid represents of the subject's annual income, and are then allowed to revise their bid. The revised bid, expressed as a percent of income, is then compared with other expenditures (for such things as clothing, rent, etc.), and subjects are allowed once again to revise their bid. Finally, subjects are told the cost of the river access program, and that their previous bid, if paid by "everyone," would be insufficient to cover the program costs, and are allowed again to revise their bid. Changes in bids at each opportunity for bid-revision are found to be statistically insignificant; final bids, however, are found to be significantly greater than initial bids (ibid., 1989, p. 689). As noted by one of the reviewers of this paper (ibid., 1989, fn.1, p. 689), however, the "significance" of these results is problematic given the process of continually "prodding" subjects used in this experiment.

The effects of "service information" (information concerning uses of a commodity) on CVM values are examined in a later paper by Bergstrom, Stoll and Randall (1990). A mail survey is used to elicit CVM values for three scenarios (concerning hunting bag levels and fish catch levels) related to the protection of wetlands along a large section of the coast of Louisiana. One set of subjects (recipients of the mail survey) was given "service information" not given to the other set: information concerning other *positive* services provided by the wetlands (wildlife encounters; natural scenery and isolation or remoteness) (ibid., 1990, p. 617). The authors find that the provision of "service information," as provided in their study, significantly increases subject's CVM values (ibid., 1990, p. 620).

What can then be concluded in terms of a response to the question of central interest here: do subjects in CVM studies base valuation decisions on well defined choice sets--do they understand the good which they are asked to value? To the extent that one requires "understanding" which is consonant with the economist's theory of value, existing evidence would seem to suggest the response NO. Significant changes in CVM values are shown to result from theoretically benign changes in the context in which the good to be valued is described: changes in the amount of time given to subjects to think about the good; changes in the way in which the good is described *vis-a-vis* substitutes and complements; and changes in the extent to which services provided by the good are described. The labile nature of CVM values with respect to changes of this nature raises (at least in our minds) serious questions as to the extent to which CVM subjects have any substantive grasp of what it is that they are asked to value in applications of the CVM.

DO CVM SUBJECTS UNDERSTAND THE HYPOTHETICAL MARKET?

Economists have long been concerned with the question as to whether or not CVM subjects understand the hypothetical market which is posited in the CVM questionnaire. Particularly important in this regard is the hypothetical payment that is requested for the contingent provision of a public good. The subjects comprehension of the hypothetical market for resulting values to reflect a real economic commitment on the subject's part. Thus, a concern is with the substance of CVM values in terms of the extent to which they reflect amounts that subjects would in fact pay if the good in question was made available to them (see, Cummings, Brookshire and Schulze, 1986, Chapter 6).[10] The basis for concern with these questions is succinctly described by Braden, *et al.* in the following way.

> "The root of the difficulties is that most of these methods (like the CVM) rely on intentions, ideals, or behavior expressed in hypothetical circumstances. Intentions are typically costless to express, or nearly so, which means that they may not be considered as carefully as are real consumption choices. Many economists are loathe to base economic values--values that will be used to allocate real resources--on information that does not grow out of real economic commitments." Braden, Kolstad and Miltz [1991, p. 12]

There have been a number of studies which have attempted to address this issue by comparing subject responses obtained within the CVM market with those derived from "markets" implicit to other methods for estimating values for public goods, usually, the hedonic price method and the travel cost method (see, as examples, Cummings, Brookshire and Schulze, 1986, Chapter 6, Cummings, Schulze and Mehr, 1978, Cummings, Schulze, Gerking and Brookshire, 1982, and Brookshire, Schulze, Thayer and d'Arge, 1982). The relevance of such comparative studies for the issue at hand is limited, however, by at least two considerations. First, such comparisons typically deal with private goods or classes of public goods which exclude those of primary interest here: environmental goods with which subjects have little experience in valuing. Indeed, in circumstances where such comparisons are possible, the case for using the CVM for value estimation is problematic (ibid., 1986, p. 163). Secondly, such comparisons beg the question that is of central importance here: that methods (and implied markets) with which CVM results are compared can themselves be shown to yield values which subjects would in fact pay (see V.K. Smith, 1986). For the purposes of our inquiry, we wish to go beyond approaches involving comparisons with other estimation methods to a review of studies which directly address the question: is there any evidence which suggests that people will actually pay amounts reported in CVM questionnaires?

One study which directly addresses this question is reported by Dickie, Fisher and Gerking, 1987. In this study, the authors obtain values for a pint of strawberries using the CVM and actually selling the strawberries to households. Their results are inconclusive. They find no significant difference in the distribution of values derived from actual sales and from the CVM. However, their price estimation equation based on CVM data performs poorly in predicting values which subjects will actually pay for strawberries.

Bishop and Heberlein (1979) find significant differences between CVM estimates for subjects' WTA for goose permits and WTA values based upon actual cash payments (*ibid.*, p. 928-929). In a later study, Bishop and Heberlein (1986) obtain CVM and actual cash values for deer permits. They found that WTP values were significantly overstated in the CVM relative to the cash market (ibid., p.131). Referring to the "familiarity" requirement (reference operating condition) for CVM values suggested by Cummings, Brookshire and Schulze (1986), Bishop and Heberlein note that "Clearly, if contingent valuation is capable of giving unbiased estimates of real values, it should have done so here." The authors note that "People were more willing to sell their goose hunting permits for real dollars than they indicated they would be in the contingent market. Preliminary results from the deer study indicate that in an auction framework, *CVM will overestimate willingness to pay . . .* money is a powerful stimulus and real money is more powerful than hypothetical money" (Bishop and Heberlein, 1979, p. 134, emphasis added).

Inferential evidence concerning the consonance of CVM values with values which subjects would pay in "real" markets is seen in the study by Seller, Stoll and Chavas (1985). In this study, CVM values are obtained for subject's willingness to pay for improvements in boat docking facilities in four lakes in Eastern Texas. Values are derived with a travel cost model, with an open-ended CVM study, and with a dichotomous choice CVM study.[11] Three results from this study are of interest here. First, both CVM applications yielded a demand relationship *with a positive slope* for one of the lakes (Lake Somerville) (ibid., p.169). The authors explanation for this anomaly is of interest for the "familiarity" issue discussed above: "It appears that specifying a contingent market under conditions where boaters were not used to paying a launch fee may have caused problems . . . This seems to add to the evidence that the contingent valuation instruments used to collect data for analysis must be designed so that behavior by the respondent is as familiar as possible" (ibid., p.169,174). Second, CVM values from their open-ended format questionnaire resulted in negative surplus measures for two of the lakes (Lake Conroe and Lake Houston). These results, along with similar results from a third lake (Lake Livingston),[12] are interpreted by the authors in the following way: they ". . . seem to indicate that people reported they were willing to pay less for an annual ramp permit *than they already paid in total launch fees over the year on a per visit basis*" (ibid., p.169, emphasis added). Third, as a test for validity or accuracy, the authors set out the hypothesis that travel cost values must be greater than CVM values. This follows from their argument that travel cost values estimate values for a total recreational experience, while their CVM format values only the boating aspects of the recreational experience. The "validity" hypothesis is rejected in two of their three tests (ibid., p.172,173).

Looking next to studies which focus on goods which are "public" in nature, Kealy, Montgomery and Dovidio (1990) examine the predictive validity of CVM values for actual cash payment for a deacidification program for lakes in the Adirondack region.[13] One group of subjects is asked their WTP for the deacidification program with the understanding that they are to actually pay their offered amount. A second group is asked for a WTP, but within a hypothetical (CVM) context. Two weeks later, this second group of subjects are asked for their actual WTP. The results suggest that ". . . individuals are more likely to overstate than to understate their (hypothetical) WTP when they are not expecting to have to make an actual payment . . . foreknowledge of an obligation to pay in accordance with one's verbal statements of willingness to pay has a positive impact on the predictive validity of contingent values. . . ." (ibid., p.259).

An experiment with a public-like good is reported by Seip and Strand (no date). 101 Norwegians were asked (in personal interviews) their (dichotomous choice: yes or no) willingness to pay 200 Norwegian kroner for membership in the Norwegian Association for the Protection of Nature (Norges Naturvernforbund, NNV), which is the largest and best established private environmental organization in Norway. 64 of the 101 subjects responded "yes." A short time later, the 64 subjects that answered yes in the CVM study were sent letters encouraging them to join the NNV at a membership cost of 200 kroner; there was no reference in these letters to the earlier CVM study. One month later, a second mailing was sent to subjects that had not joined the NVA as a result of the first letter; again, reference was not made to the CVM study. At the end of the second mailing, only six of the 64 "yes" respondents in the CVM had actually paid the 200 kroner to join the NVA.

A sample of 25 of the 58 "yes" respondents in the CVM study who had not responded to invitations to do what they had said that they would do were interviewed by telephone. Emphasizing the scientific nature of their study, the authors reminded the subjects of their "yes" response to the CVM questionnaire, and the two mailings which offered them the opportunity to do what they had said that they would do. The subjects were asked for reasons underlying their lack of response to invitations to join the NNV. 24 of the 25 subjects indicated that their willingness to pay expressed in the CVM study *was an expression of their willingness to pay for environmental goods in general, not their willingness to pay for the NNV in particular*! When asked if they would like to change the WTP offered by them in the original CVM study, 17 of the 25 subjects indicated that they would lower their response. Seip and Strand conclude that the results are ". . . discouraging, by indicating that the CVM can imply quite serious biases of overvaluation, in particular when (like here) the good to be valued is rather abstract and it is difficult to attach concrete environmental values to it" (ibid., p.3).

Based upon these empirical results, there would appear to be a total lack of consonance between values obtained within the hypothetical "market" of the CVM and values obtained from non-hypothetical markets. The position that CVM subjects understand the hypothetical market would then be difficult to justify. It would seem to be clear, however, hypothetical payment is quite a different thing from real payment. Bishop and Heberlein's conclusion is particularly telling in these regards: "[M]oney is a powerful stimulus and real money is more powerful than hypothetical money" (1979, p.134).

WHAT IS THE EMPIRICAL CONTENT OF CVM VALUE RESPONSES?

The focus of this section turns to a set of issues which are more technical in nature than those discussed above, but which are extremely important. There appears to be a general skepticism, at least casually, among researchers that at least some values reported by CV survey respondents are untruthful or unreasonable. As a consequence, doubt is cast on the validity and reliability of conclusions drawn from empirical analysis based on reported values. The way in which supposed "untruthful" or "unreasonable" (most often referred to as "protest" and "outlier" bids) values can substantially affect CVM value estimates involves removing zero bids and "large" bids from the sample.

Reviewing methods currently used to deal with this issue is intended to reflect the general lack of theory-based rationale in these methods.

Field data obtained with the CVM are characterized as suffering from biases and errors that derive from such sources as poor instrument design, measurement error or the lack of incentives for the respondents to seriously consider their answers or respond truthfully. While some of these problems are not unique to contingent value studies, others, resulting from the hypothetical nature of the market, deserve special attention. Some problems can affect the accuracy, or unbiasedness, of any calculated statistic. Others can affect the reliability, defined as the variance (or, more generally, the mean squared error) of the estimate (Kealy, Montgomery and Dovidio, 1990).

Consider what is a standard procedure presently employed for analyzing field CV data. The sample data are collected and some measure of the average value is calculated. This may be as simple as calculating the sample mean or median, or as complicated as estimating the expected mean bid from a multivariate regression model in which individual or household characteristics are included as "explanatory" variables. Once an average value has been obtained it is applied to the relevant population (say, the entire population of the United States,) to obtain a total value that can then be used in a comparative policy analysis. Errors in collecting the data, calculating the average and imputing a total can be multiplied throughout the process described above such that the final figure is no better than a random guess.

Without undertaking a lengthy summary and discussion of the vast literature that addresses the multitude of biases that have been identified when CVM data are analyzed via the procedures described above (see Mitchell and Carson, 1989). The focus will be on the specific issue of the treatment of outliers and protest bids, their sources and possible remedies for the problems they may cause for the estimation process. Outliers are values at the extremes of the distribution. Extreme values, while theoretically possible for most distributions (for example, the normal distribution is defined over the entire real number line), are usually associated with measurement error or coding error. When interpreted this way, outliers are random in the sense that there would appear to be no way to identify them other than by some arbitrary rule. Another interpretation of outliers is that they are response errors. That is, the respondent may give a value that is implausible by miss-speaking or not "thinking" about the response. For example, a respondent may state a value that is larger than their annual income or wealth. Such an outlier can be identified by certain cross-checks on other data collected from the respondent and "edited" out (Carson, 1987, p.4). Basically, some "rule" must be used to identify outliers.

Protest bids are responses to valuation questions that are not valid bids. They are often referred to in conjunction with missing values as "protest bids" (Schulze, McCelland and Waldman, 1991). However, it is our contention that missing values should be treated separately from the class of protest bids that take on non-negative values since it is these responses that have the greatest potential for confusing the calculation of the sample statistics. Protest bids are defined as responses that reflect a person's feelings about the subject ("I don't care about protecting wildlife"), the institution ("I won't pay higher taxes because the government wastes that money anyway"), or the instrument ("I never fill out mail surveys") rather than indicate the person's true valuation. As such, protest bids could take any value, but it is the conventional wisdom that they are most likely to be zeros. Protest bids, like outliers, must be identified by some "rule", possibly related to answers given by the respondent to follow-up questions.

Related to outliers and protest bids are strategic bids. The notion of strategic bids, and strategic behavior, has as its origin the work of Samuelson (1954) on the efficient provision of public goods. The idea is that people, when asked for their valuation of a marginal unit of a public good, will have an incentive to misrepresent their preferences, depending upon the method and value of the amount they will be asked to contribute to the provision of the

good.   Brookshire, et al. (1976) present the standard argument.   In it, an
individual who believes that the level of provision will be determined by the
mean response will have an incentive to overrepresent their true value if that
true value is larger than the mean, and have an incentive to underrepresent if
their true value is below the mean. The consequence of this behavior, according
to Brookshire, et al., is that strategic behavior will cause more small and large
bids compared to what would be observed in the absence of strategic behavior.
It is this consequence that relates strategic bidding to outliers and protest bids
since these phenomena generally occur at the extremes of the bid distribution.
While no formal test exists to identify strategic behavior, and despite assertions
that is it not a significant problem (Mitchell and Carson, 1989, p.153), it is
apparent that, if strategic behavior manifests itself as extreme bid values, then
it may be possible to identify strategic, (and therefore, invalid,) bids through a
"rule" or model that describes such behavior.

Outliers and protest bids can have a significant effect on the size of
sample statistics calculated with the data collected in a CV survey. According
to the arguments above, this effect, or bias as it has come to be known, should
be removed from the calculations or else policy conclusions will be based on
statistics that do not have the desirable unbiased and efficient characteristics.
There exists a number of methods, or "rules", that have been developed to
identify, or compensate for, outliers and protest bids. Protest zeros, often the
only type of protest bid considered by researchers, are usually identified by the
response to a question that is triggered in the questionnaire by a zero bid. If the
respondent gives a zero value for his or her WTP then the person is asked if that
response represents a valid zero response or something else. It the answer is
"something else", the respondent is generally identified as a protestor and the
data point removed from the calculations. There is, as far as we know, no
reported attempt to identify, separately, non-zero protest bids in any field CV
study.

Outliers, on the other hand, have been treated in a non-behavioral way in the literature. That is, standard robust estimation solutions have been applied to the data to compensate for the effect of outliers rather than attempt to model them explicitly, or at least link them to other observable data. Statistical analysis designed to adjust for outliers focus on non-parametric analysis, or order statistics such as the median. One method currently receiving attention is the calculation of α-trimmed means (Mitchell and Carson, 1989, p.226). This method weights the αxn smallest and αxn largest observations in the data set by zero in the calculation of the sample mean (n is the sample size). The resulting mean is the α-trimmed mean and is smaller in value than the regular, full sample, mean when the bid distribution is skewed to the right. Large and small (generally some zero) values are "removed" by this method, thereby addressing the issue of outliers, albeit in a purely statistical way. The trimmed mean will, in general, have a smaller mean square error than the mean in the presence of outliers if the underlying distribution is normal, however, this property does not carry over to the case when the distribution is non-normal, as the distribution of observed bids would seem to suggest (DeGroot, 1975, p.478).

The problem with trimming observations from the observed distribution is that the method is not related to any hypothesized or observed behavior of the respondent. For example, the trimming procedure will "eliminate" some of the zero bids, but are these bids protest zeros as identified by the protest question? Should the researcher use the protest question to eliminate protest zeros but use the trimming device to eliminate large bids, and can these methods coexist?

Another, alternative, method that acknowledges outliers is the Box-Cox method proposed by Schulze, McClelland and Waldman (SMW) [1991]. Responding to the observation that bid distributions appear skewed to the right and very non-normal, SMW suggest estimating a bid function using the Box-Cox transformation as opposed to the standard linear transformation used in ordinary least squares. Essentially this method transforms the data such that the resulting residuals are distributed normally, which would not be the case if the transformation were not done. Implicit in this method is the assumption that outliers are not errors or observations that do not belong to the underlying bid distribution, but rather that the bid distribution is non-normal. SMW propose this method as an alternative to trimming which they criticize as giving results that are dependent upon the choice of the α level, a valid criticism. SMW do not present the Box-Cox method as a way of adjusting for protest bids or protest zeros, however.

While the methods discussed above indicate a recognition on the part of CV researchers of the serious problem of outlier and protest bids, the present state of the art involves little more than arbitrary choices for what the researcher takes to be a "reasonable" or an "unreasonable" value.[14]

CONCLUDING REMARKS

In this paper we have addressed the question as to the extent to which there exists empirical evidence which would support the presumption that CVM values for any given public good are in some substantive way consonant with subjects' true values for the good. The literature relevant for assessments of this question is clearly not robust. Relatively few dimensions of the "true value" question have been investigated, and there are few studies which focus on any of these dimensions. Most studies which have been published to date are exploratory in nature, and raise more questions than they answer. As but one example in this regard, the Bergstrom, Stoll and Randall [1990] study of the effects of "service" information on CVM values demonstrates that some types of positive information can effect CVM values. Unanswered by the study are the implied questions: when does adding information cease to inflate values; what kinds of information should be provided to subjects; and how might negative service information affect values?

There then exists no basis for definitive conclusions as to the substance of CVM values. The evidence which does exist, however, would appear to reject any presumption that CVM values may be interpreted as reflecting subjects' real values for environmental goods, "real" in the sense of reflecting a value which a subject would in fact pay for the environmental good in question. CVM values appear to be extraordinarily sensitive to reasonable changes in the context in which the good is described; it is simply not at all clear that subjects understand the good which is to be valued with the CVM. Divergences between what people <u>say</u> they will pay for a good within the context of the CVM's hypothetical market and what they <u>actually</u> pay for a good leaves serious questions as to the extent to which they really understand the hypothetical market. Until the CVM can be structured in a manner in which subjects demonstrably understand these key components of any valuation process, results from applications of the CVM are ill-described as "values."

NOTES

1.   A version of this paper was presented by R. Cummings at the Department of Mineral Economic's 1990 John M. Olin Distinguished Lectureship in Mineral Economics, held at the Colorado School of Mines on November 9, 1990.

2.   See U.S. Department of the Interior, 1986. Also, see Smith, this volume for a general discussion of Type A and Type B damages.

3.   See, for example, Phillips and Zeckhauser, 1989.

4.   Efforts to model consumer behavior where information and time conditions may parallel those found in most applications of the CVM are seen in Hoehn and Randall, 1987. The critical hypothetical conditions of the CVM "market," however, are not included in these efforts.

5.   The word "compared" admits of indifference. Thus the agent could well say that given his current knowledge of the two bundles he does not prefer one to another. This statement in perfectly consistent with the Completeness axiom.

6.   To non-economists this axiom sounds trivial and silly, but it is needed to ensure that certain logical properties of the system of preferences are satisfied. If it were not true the axiom system would admit contradictions at every step of virtually every proof.

7.   Commodities that the agent does not like are called "bads". In most practical applications one can just reverse the directions of axes and measurement to ensure that we are always talking about commodities that are "goods" in the sense defined by the Monotonicity axiom.

8.   A common way to state this axiom is that the indifference curves (the level sets of the utility function) have superior sets (the consumption bundles preferred to those on the indifference curve) that are convex. In more homely terms, this axiom ensures that indifference curves are bowed into the origin.

9.   A recent study by Kahneman and Knetsch (1992) addresses issues related to those discussed above. These results are not discussed here given the controversy that surround the reliability of data obtained by these authors. See, Burness, Cummings, Ganderton and Harrison (1991), V.K. Smith (1992) and Harrison (1992).

10.   For a particularly bombastic critique of the CVM in these regards see phillips and Zeckhauser, 1989.

11.   An open-ended format solicits a dollar bid from the respondent based upon a given dollar figure.

12.   Surplus was not calculated for the fourth lake, Lake Sommerville, because "(...) the demand curve was not downward sloping, lay in the fourth quadrant and was considered unreliable." Sellers, Stoll and Chavas (1985, ft. p. 171).

13.   The strength of these authors' conclusions is weakened by small sample sizes. For example, the authors note that "The reader should be aware, however, that because some of the cells in Table VI have less than five observations, these test results must be viewed with some caution." Kealy, Montgomery and Dovidio (1990, p. 259).

14.   A quite different characterization of the outlier and protest bid problem is based on the assumption that the observed distribution of bids is a mixture of more than one distribution. There is a basic underlying, distribution of non-negative, valid and truthful bids. But confronted with this distribution are a number of other bid distribution, for example, one that describes extreme large bids, another that describes protest bids, many of which are zero. The problem is to separately identify which of these distributions, or, less ambitiously, identify which distribution each observation belongs to. this may even be too difficult to do, so at the very least we would like to be able to identify which observations belong to the distribution of valid bids. An attempt at this type of analysis described would involve the following. Take three distributions which are hypothesized to describe the observed data, one for the valid bids, and one each for the outlier and protest bids. Prior probabilities that any observation belongs to any distribution are established, based on experience of other CV data, focus group discussions or instrument pretesting (the Idea that researchers have priors regarding many aspects of empirical analysis in economics is not new or unfamiliar).   A random sampling of all possible assignments of observations to processes is used in a Monte Carlo study of the posterior distribution of the mean and standard deviation of the valid bid distribution. The method follows that of Box and Tiao [1968]. While this type of analysis is in its early development stages, it is theoretically appealing due to its explicit recognition of the processed generating outliers and protest bids.

REFERENCES

Bergstrom, John C., John R. Stoll and Alan Randall, "Information Effects in Contingent Markets," Amer. J. Agricultural Economics, 71(3), 687-691, August, 1989.

Bergstrom, John C., John R. Stoll and Alan Randall, "The Impact of Information on Environmental Commodity Valuation Decisions," Amer. J. Agricultural Economics, 72(3), 614-621, August, 1990.

Bishop, Richard C. and Thomas A. Heberlein, "Measuring Values of Extramarket Goods: Are Indirect Measures Biased?" Amer. J. Agricultural Econ., 61, 926-930, December, 1979.

Bishop, Richard C., and Thomas A. Heberlein, "Does Contingent Valuation Work?", Chapter 9 in Cummings, Ronald G., David S. Brookshire, and William D. Schulze, Valuing Environmental Goods: An Assessment of the Contingent Valuation Method, Roman and Allanheld (Totowa, N.J.), 1986.

Box, G. E. P. and Tiao, G. C. "A Bayesian Approach to Some Outlier Problems" Biometrika, Vol55 No.1 1968 119-129.

Braden, John B., Charles D. Kolstad and David Miltz, "Introduction," chapter 1 in Braden, John B. and Charles D. Kolstad (ed.s), Measuring the Demand for Environmental Quality, North-Holland (New York), 1991.

Brookshire, D.S., B.C. Ives and W.D. Schulze, "The Valuation of Aesthetic Preferences," J. Environ. Econ. Mngt. 3(4), 325-346, December, 1976.

Burness, H.S., Ronald G. Cummings, Philip Ganderton, and Glenn W. Harrison, "Valuing Environmental Goods: A Critical Appraisal of the State of the Art", Chapter 22 in Dinar, Ariel, and David Zilberman (ed.s), The Economics and Management of Water and Drainage in Agriculture, Kluwer Academic Publishers (Boston, 1991).

Carson, Richard T., "Constructed Markets," chapter V in Braden, J.B. and C.D. Kolstad (ed.s), Measuring the Demand for Environmental Quality, North-Holland (New York), 1991.

Cummings, R.G., W.D. Schulze, S.D. Gerking and D.S. Brookshire, "Measuring the Elasticity of Substitution of Wages for Municipal Infrastructure: A Comparison of the Survey and Wage Hedonic Approach," J. Environ. Econ. Mngt., ...

Cummings, R.G., W.D. Schulze and A.F. Mehr," Optimal Municipal Investment in Boomtowns: An Empirical Analysis," J. Environ. Econ. Mngt., 5, 1978.

Cummings, Ronald G., David S. Brookshire, and William D. Schulze, Valuing Environmental Goods: An Assessment of the Contingent Valuation Method, Roman and Allanheld (Totowa, N.J.), 1986.

DeGroot, M. Probability and Statistics, Addison Wesley (Mass. 1975).

Dickie, Mark, Ann Fisher and Shelby Gerking, "Market Transactions and Hypothetical Demand Data: A Comparative Study," J. Amer. Statistical Assn., 82(397), 69-75, 1987.

Freeman, A. Myrick III, "On Assessing the State of the Arts of the Contingent Valuation Method of Valuing Environmental Changes," Chapter 10 in Cummings, Ronald G., David S. Brookshire, and William D. Schulze, Valuing Environmental Goods: An Assessment of the Contingent Valuation Method, Roman and Allanheld (Totowa, N.J.), 1986.

Harrison, Glenn W., "Valuing Public Goods With the Contingent Valuation Method: A Critique of Kahneman and Knetsch," J. Environmental Econ. Mngt., forthcoming, 1992.

Hoehn, John P., "Valuing the Multidimensional Impacts of Environmental Policy: Theory and Methods, Amer. J. Agricultural Econ., 289-299, May 1991.

Hoehn and Loomis 1991 The Prospect for Complementarity in Valuing Intra-Regional Environmental Programs, unpublished ms., 16 pp., Department of Agricultural Economics, Michigan State University, East Lansing, at p. 2.

Hoehn, J.P. and A. Randall, "A Satisfactory Benefit-Cost Indicator from Contingent Valuation," J. Environ. Econ. Mngt, 14, 226-247, 1987.

Hoehn, J.P. and A. Randall, "Too Many Proposals Pass the Benefit-Cost Test," Amer. Econ. Rev., 79(3), 544-551, June, 1989.

Kealy, Mary Jo, Mark Montgomery and John F. Dovidio, "Reliability and Predictive Validity of Contingent Values: Does the Nature of the Good Matter?" J. Environ. Econ. Mngt., 19, 244-263, November, 1990.

Mitchell, R. C. and Carson, R. T. Using Surveys to Value Public Goods, Washington: Resources for the Future, 1989.

Phillips, Carl V., and Richard J. Zeckhauser, "Contingent Valuation of Damage to Natural Resources: How Accurate? How Appropriate?", Toxic Law Reporter, 520-529, 1989.

Samuelson, P. A. "The Pure Theory of Public Expenditures", Review of Economics and Statistics, Vol 36, 1954 387-389.

Schulze, W, McClelland, G. and Waldman, D. "Contingent Valuation Methods and the Valuation of Environmental Resources", presented at the International Conference: Economy and the Environment in the 1990's Switzerland, Aug 1990.

Seip, Kalle and Jon Strand, "Willingness to Pay for Environmental Goods in Norway: A Contingent Valuation Study With Real Payment," mimeo 25 pp., Department of Economics, SAF Center for Applied Research, University of Oslo, Norway (no date).

Sellar, C., J.R. Stoll and J.P. Chavas, "Valuation of Empirical Measures of Welfare Change: A Comparison of Nonmarket Techniques," Land Econ., 61, 156-175, 1985.

Smith, V. Kerry, "To Keep or Toss the Contingent Valuation Method," Chapter 11 in Cummings, R.G., D.S. Brookshire and W.D. Schulze (ed.s), Valuing Environmental Goods, Rowman and Allanheld (Totowa, N.J.), 1986.

Smith, V. Kerry and W. Desvousges, "An Empirical Analysis of the Economic Value of Risk Changes," J. Political Economy, 95(1), 89-114, February, 1987.

Smith, V. Kerry, "Arbitrary Values, Good Causes and Premature Verdicts," J. Environmental Econ. Mngt., 22(1), January, 1992.

Tolley G.S., et al., "Establishing and Valuing the Effects on Improved Visibility in the Eastern United States," Report to the U.S. EPA, Washington, D.C., 1984.

U.S. Department of the Interior, "Natural Resource Damage Assessments: Final Rule", 51 FR 1148, 27673-27753, August 1, 1986.

CHAPTER IV

INTERGENERATIONAL FAIRNESS & GLOBAL WARMING

Global
Q25

D 91

Ralph C. d'Arge

INTRODUCTION

For the second time in the twentieth century, the first being the introduction of nuclear weapons and the possibility of a "nuclear winter", mankind is confronted by the possibility of non-sustainability due to climate change. That is, continued tropospheric warming due to the burning of fossil fuels is likely to place the global climate outside of the limits of recorded history. Whether human adaptation is great enough in modern societies to offset substantially warmer climates is, at least, debatable. Certainly, the goals of sustainability outlined by Kneese (1990) of "continued well-being of people in developed countries, improvement in the well-being of people in developing countries and protection and maintenance of a safe and attractive environment", become highly questionable.

There are at least three ideas with respect to global warming that need to be explored. The first is to examine the evidence on the hypothesis that global warming will have substantially different regional impacts, with some regions gaining and others being significant losers. The second to be evaluated is the extent and speed at which the climate warming will occur. According to recent calculations, past burning of fossil fuels has led to substantial warming potential for the near future and present emissions of carbon dioxide (CO_2) are accelerating the potential for greater warming. Average global temperature is forecast to potentially change by as much as 30 percent, based upon celsius measures, from present temperatures. The third idea to be examined is, if the other two are correct, the efficient and appropriate set of actions for the multiple set of global societies to take now.

REGIONAL IMPACTS OF GLOBAL WARMING

One of the main sources for predicting the impacts of global warming is climate simulation modelling using General Circulation Models (GCMs). GCMs typically solve 200,000 equations on each run and are extremely complex. Yet, the results from different GCMs often agree well with each other, and with historical temperature data, over large scales (global, hemispherical, zonal). More than 100 independent studies have given estimates of average global warming with the 1.5° to 4.5° Celsius range for a scenario indicating a doubling of CO_2, with values near 3°C tending to be favored (MacDonald 1988 p. 437).

As results of the GCMs are examined on successively smaller scales, eventually focusing on subcontinental regions, significant differences arise between models. GCMs treat the world as uniform blocks (at their most detailed based on a 500 Km grid size) ignoring topographical and meteorological variations within these blocks. Also, the use of averages, when concern is for scales below the regional extent for which the average is estimated, can be misleading. Two models can have precisely the same average for a regional variable but still give significant spatial differences for that variable within the region. Thus, GCMs are not ready to be used for quantitative prediction even at the level of a multi-state region, let alone a particular state, county, or city (Grotch 1988 p. 253). Therefore, the following is a qualitative picture of the regional effects of global warming based on broad latitudinal zones.[1] In addition, we identify specific areas as being particularly susceptible to the expected adverse effects of global warming.

The regional temperature and precipitation changes due to climate change via the greenhouse effect are summarized in Table 1. The temperature changes, presented as multiples of the global and annual average temperature changes, are taken from computer modelling results for the northern Hemisphere. A qualitative assessment of precipitation changes is given, but is supported by a number of studies. The general picture is one of far greater annual average changes in temperature toward the poles and variable interregional changes.

In the Northern Hemisphere, above latitude 60° north, temperature change is predicted at two to two and half times faster during the winter months and greater than the global average. Withdrawal of summer ice pack could leave the Arctic ice-free around Spitzbergen and along the north Siberian coast. A positive feedback occurs as the duration of snow cover is reduced affecting albedo and causing further warming. Increased precipitation in the high-latitude regions of the Northern Hemisphere leads to greater runoff into the Arctic Basin. The growth of Borcal forests will be stimulated and the rate of decay of organic material enhanced. Most effects in the region carry both benefits and costs, so that neither net benefits nor net costs can be identified as potentially dominant. For example, while opportunities for increased use of the Northeast and Northwest passages should be beneficial, there are potentially serious effects on national and international security since the poles would be more easily accessed.

The mid-latitude zone, from 30° to 60°, is anticipated to warm more than the global average but less than the high latitudes. Winter warming is greater than summer, while summer rainfall decreases. The main impact is expected to be on unmanaged ecosystems, forests, and agriculture. The reproductive success of many tree species would be reduced and both tree and plant mortality would increase. The extent and timing of forest dieback depends on the estimated sensitivity of the climatic response and actions taken to control

Table 1: Regional Scenario for Climate Change

Region	Temperature Change (as a multiple of global average)		Precipitation Change
	Summer	Winter	
High Latitudes (60°-90°)	0.5x - 0.7x	2.0x - 2.4x	Enhanced in Winter
Mid Latitudes (30°-60°)	0.8x - 1.0x	1.2x - 1.4x	Possibly reduced in summer
Low Latitudes (0°-30°)	0.9x - 0.7x	0.9x - 0.7x	Enhanced in humid zones with heavy current rainfall

Source: World Meteorological Organization (1988) Table 1

greenhouse gas emissions. Major effects on forests could begin between 2000 and 2050, or given serious global restrictions on gas emissions be pushed back to 2100. A faster rate of temperature change worsens the expected forest damage.

Global food supplies, as a whole, should be maintained via adaptation based on agricultural research, except in the case of rapid warming. Negative effects on productivity in the lower-latitudes (around 30°), due to greater evapotranspiration, may be balanced by positive effects in the higher latitudes (around 60°), due to a longer growing season. However, irrigated agriculture in the semi-arid areas of the mid-latitudes will probably be adversely affected, unless substantial offsetting irrigation developments are implemented.

The semi-arid tropical regions, from 5° to 35° north and south, already suffer from unevenly distributed seasonal average precipitation with high spatial and interannual variability. This latitudinal belt as a whole is expected to suffer reduced precipitation. There has been a pronounced downward trend in precipitation since the early 1950's for this region which has resulted in prolonged drought and aggravating the on-going desertification process. Future climatic changes could worsen these current problems.

Global warming leads to a 5 to 20 percent increase in rainfall in humid tropical regions. Water levels along coasts and rivers are increased by sea level rise, greater frequency of tropical storm surges, and rising peak runoff. Large areas of flooding and salinization are likely to result. In addition, the spatial and temporal distribution of temperature and precipitation could change.

The outline of potential regional effects of global warming, given above, implies numerous positive and negative impacts. The evidence shows that climatic change will have different impacts for different regions and countries. Global warming would alter precipitation patterns, evapotranspiration, and the length of growing seasons, affecting agriculture and forestry. It is also likely to increase the rate of sea level rise threatening capital values in low lying areas. Thus, as has been previously recognized (see d'Arge et al. 1982, Kosobud and Daly 1984, Broadus 1986, Barbier 1989, and Glantz 1990) the benefits or costs of induced climatic change are likely to be unevenly distributed between nations. In particular, sea level rise and changes in semi-arid regions imply greater costs for the developing nations of the world.

Sea level rise is already of concern with one half of the global population inhabiting coastal regions, while average sea level increased at 0.01 meters per decade during the last century. An acceleration of this rate would occur as global warming melted land ice and caused thermal expansion of sea water. Estimates of sea level rise vary but a probable scenario is 0.9 to 1.7 meters by 2100 (Thomas, 1986 and Tutus, 1986). Jaeger (1989) describes typical impacts as being beach and coastal margin erosion, loss of wetlands, increased frequency and severity of flooding, and damage to coastal structures and water management systems.

Low lying developed countries already possess sea defenses which could be modified to account for the dangers of a 1 meter sea level rise. Goemans (1986) estimated the cost of protecting against such a rise for the Netherlands, where extensive coastal protection already exists, at $4.4 billion. In the United States the cost of maintaining shores under threat on the East coast is in the range of $10 to $100 billion, for a 1 meter rise (Jaeger 1989 p. 101). Park *et al.* (1986) have predicted a loss of 40 to 73 percent of U.S. coastal wetlands by 2100. This might be reduced to 22 to 56 per cent if new wetlands are allowed to form on currently inland areas.

Developing countries would experience serious impacts by a 1 meter sea level rise or less. Flood disasters are particularly threatening for the delta regions of South Asia and Egypt. Egypt would have 12 to 15 percent of its arable land area flooded, affecting 7.7 million people, 16 percent of its' population (Broadus 1986). While in Bangladesh 8.5 million people would be affected, 9 percent of the population, and 11.5 percent of the land area would be flooded (Broadus 1986). The 177,000 Maldivians are in a situation typical of many low lying island populations. They could be totally displaced by even a small sea level rise due to the encroachment of storm surges. The Maldives are coral islands on average less than a meter and a half above sea level (Economist 1988 p. 36).

The ability of nations to adapt to sea level rise is determined by their capital resources and technical knowledge. Developed nations are, therefore, better equipped to respond, at least initially, to predicted sea level increases. Many less developed countries will find their current floods becoming worse with a direct impact in terms of human suffering, permanent dislocation, and loss of life. As Jaeger (1989 p. 101) has stated, "In developed countries, lowland protection against sea level rise will be costly. In developing countries without adequate technical and capital resources it may be impossible".

Another area in which developing countries stand to suffer disproportionately is the semi-arid regions. The important arid zones of the world are found around latitude 30° north and south, where most of the large deserts are situated; e.g., the Sahara (N. Africa), the Syrian desert and Arabia (West Asia), Death Valley and surrounding areas (U.S.), and the Australian deserts (Waller 1966). Only eleven nations are wholly arid while twenty-seven nations have some territory within the arid zone but the larger part of their land area outside of it (see Table 2). The peripherally arid nations are those in this latter group for which aridity is significant only at the regional level.

Characteristically, the equilibrium of water, soil, geological, and vegetative processes in arid lands is a delicate one, so that a slight shift in one aspect may initiate a disastrous chain of events (White 1966 p. 15). The scarcity and variability of rainfall are dominant elements in the complex of physical factors affecting arid lands. The natural limits of arid lands are set by climatic conditions influencing the surplus and deficit of water for plant growth. The swing from crop success to utter failure is quick and frequent and the greatest risk is in semi-arid areas. Except for the U.S. and the desert oil exporting nations, most arid or semi-arid nations have low levels of capital foundation and/or technical expertise required to develop strategies to protect against drought or extreme variations in climates.

High temperatures over long periods increase the energy available for evapotraspriation so that the potential water loss is greater than the amount available. Such high temperatures can intensify the chemical processes in a plant causing death, i.e., hot spots. Extreme temperature may lead to crop failure and stress both animals and humans (Waller 1966).

An increase in frequency, duration, and severity of droughts in the Great Plains of the U.S. would lead to a more rapid depletion of the Ogallala aquifer (partial source of groundwater to eight western states). The semi-arid western and mid-western states in the U.S. may be particularly susceptible to the effects of climate change on river systems (d'Arge 1975). Large decreases in surface runoff (40 to 76 percent) could accompany a 2°C temperature increase (Barbier 1989 p. 24). The effect of a CO_2 doubling on U.S. agriculture in the western states has been estimated by Adams, et al. (1988) to induce crop losses of $6 billion to $33 billion (1982 dollars). However, almost half of the consumers' surplus losses from climate change in the U.S. fall on foreign consumers.

Adams, et al. (1988 p. 348) also note that climate change is not a food

Table 2:　The Arid Nations

Description	Number	Percentage of Nation Arid or Semi Arid	Nations
Core	11	100	Bahrain, Djibouti, Egypt, Kuwait, Mawitania, Oman, Quater, United Arab Emirates, Saudi Arabia, Samalik, South Yemen.
Predominately Arid	23	75 - 99	Afghanistan, Algeria Australia, Botswana, Cape Verde, Chad, Iran, Iraq, Israel, Jordan, Kenya, Libya, Mali, Morocco, Namibia, Niger, North Yemen, Pakistan, Senegal, Sudan, Syria, Turisia, Upper Volta.
Substantially Arid	5	50 - 74	Argentina, Ethiopia, Mongolia, South Africa, Turkey.
Semi Arid	9	25 - 49	Angola, Bolivia, Chile, China, India, Mexico, Tanzania, Togo, U.S.A.
Peripherally Arid	18	< 25	Berin, Brazil, Canada, Central African Republic, Ecuador, Ghana, Lebanon, Lesotho, Madagascar, Mozambique, Nigeria, Paraguay, Peru, Sri Lanka, USSR, Venezuela, Sambia, Zimbabwe.

Source:　Heathcote (1983) p. 9.

security issue for the U.S. Even in the most extreme case their analyses indicate that the productive capacity of U.S. agriculture will be maintained at a level that avoids major disruptions to the supply of the modelled commodities. Consumers in the U.S. face slight to moderate price rises under most scenarios, but supplies are adequate to meet current and projected domestic demand. Exports, however, experience major reductions. Exported commodities in some scenarios decline by up to 70 percent, assuming constant export demand.

While arid regions of North America may suffer as climate approximates a double CO_2 scenario the northern latitudes may simultaneously benefit from an increase in the length of the growing season and a shift northward and eastward in the agricultural belt. Agricultural adjustments could increase yields via the extension of winter wheat into Canada, a switch from hard to soft wheat in the Pacific Northwest due to increased precipitation, and an expansion of areas in fall-sown spring wheat in the southern latitudes due to higher winter temperatures (Rosenzweig 1985 p. 380). Kelejian and Varichek (1982) believe the U.S. to be slightly below optimal temperature for wheat and corn so that a small temperature increase (less than a double CO_2 climate) could increase U.S. yields by 5 percent. If the rest of the worlds production fell, as they predict, the position of U.S. agricultural producers would improve through exports. However, more recent evidence shows other countries may also benefit.

Higher temperatures tend to favor yields of cereal crops in regions where temperature limits the growing season. For example, in the central European region of the former USSR wheat yields have been projected to increase a third under a double CO_2 scenario (Parry and Carter 1986 p. 275). Smit, et al. (1988) have reviewed several recent studies and concluded that current evidence suggests a warmer climate could create a more favorable environment for wheat and grain corn in Canada, Northern Europe, and the USSR.

In the case of semi-arid regions of less developed countries the impact of climate on economic activities is most severe in the agricultural sector of the desert and areas marginal to it, e.g., in the Sudan and Sahel. Areas marginal to the desert can give a false sense of security to their inhabitants during periods of successively wet years, but are highly vulnerable in periods of drought (Oguntoyinbo and Odingo 1979 p. 2). Continued wet-year land use and livestock practices during droughts is a contributing factor to desertification.

Moderate, or worse, desertification has occurred in 80 percent of the agricultural lands of the worlds arid regions. Irrigated lands suffer waterlogging and salinization, grazing lands lose plant cover, and rainfed croplands are desertified by soil erosion. Those areas currently undergoing severe desertification are agricultural economies with the livelihood of their people being 27 percent urban based, 51 percent cropping based, and 22 percent animal based (Dregne 1983 p. 20).

The population of the arid regions is approximately 700 million with 78 million living in areas where severe desertification has occurred. Among these people, 50 million have been estimated as already suffering a loss in ability to support themselves and are under pressure to migrate to overcrowded cities. The population of the moderately desertified portion of the arid lands is at least eight times that of the severely affected regions (Dregne 1983 p. 19-21).

Climatic changes due to the greenhouse effect will be imposed upon the already fragile ecosystems of these semi-arid regions. The prediction of reduced precipitation for the latitudinal zone from 5° to 35° north and south would, alone, aggravate the existing climatic problems of the arid regions. In addition, the temperature increase can be expected to add to crop and livestock stress and fatality. The size and subsistence level of population living in this region implies the possibility of a far greater cost than expected shifts northward of agricultural production predicted for North America and Europe.

The failure of crops in one region may be balanced by productivity increases in other regions, as mentioned earlier for the middle latitudes. Newman and Pickett (1974) have suggested that generally an equilibrium climate obeys a law of conservation, born out by the history of global climatic variations. That is, whenever one large area gets too little precipitation, another gets too much. Thus, one strategy to respond to the problems of arid lands is to organize resource use to compensate for the inevitable shortages of production in one location by transfers of surpluses from other locations.

Global impacts of greenhouse warming on agriculture will be unevenly distributed with the North-South divide playing an important role in the ability to adapt. The major Northern Hemisphere food-exporting, and capital intensive, countries are in a better position than the low-income, food-importing developing countries. The wealthier northern nations while facing substantial changes in types of crops cultivated, water distribution and areas of available land, are more able to adapt. These countries "...tend to have a surplus of land available for production as well as accumulated stocks of some produce, highly developed agricultural R & D infrastructure and techniques, efficient marketing credit and information systems and extensive water management and control systems." (Barbier 1989 p. 25) If adaptation raises costs and reduces supplies in the food-exporting North, the food-importing South is likely to suffer the most.

There are 65 low-income, food-deficit, countries with the greatest risk group being those who have failed to increase per capita food production in the past. Unfortunately, most of this latter group have extensive agricultural areas located in arid and semi-arid regions. As a result global climate change threatens to make a bad situation worse. Currently there are excess supplies of world food production but 730 million people in developing countries are denied enough energy from their diet to allow them an active working life (Barbier 1989 p. 26).

INTERTEMPORAL IMPACTS OF GLOBAL WARMING

Mean global temperature has been much warmer in the past than at present; during the Holocene climatic optimum (5000 to 6000 B.P.) 1°C warmer, during the last interglacial warming (125,000 B.P.) 2°C higher, and during the Pliocence (3 to 4 million years B.P.) 3°C to 4°C higher (McDonald 1988). However, during the last 10,000 years, from the warm Holocene to the Little Ice Age, the mean temperature of the northern hemisphere varied by no more than about 2°C (Gates 1983). In recent years, mean global temperature has been about 15°C. Thus, mean global temperature has varied by less than 14 percent over this time.

The greenhouse effect represents a potentially drastic temperature increase over a relatively short space of time. The earth's surface temperature has increased between 0.5° and 0.7°C (3-5 percent) since 1860 (Abrahamson 1989 p. 10). Hansen, et al. (1986) estimate the warming at most mid-latitude northern hemisphere land areas would be 0.5°C to 1.0°C by 1990 to 2000, and 1°C to 2°C by 2010 to 2020. Toward the end of the next century, the planet could warm up by 4.5°C (or a 30 percent increase) (Smorgorinsky 1983). Within 100 years the earth could be significantly warmer than it has been in several million years and certainly beyond anything observed during recorded history. Such changes are well beyond known human capabilities of adaptation and raise fundamental questions regarding what constitute limits to sustainability.

The expected consequences of sea level rise and agricultural and forestry damages suggest an overall negative impact due to a 2°C or more increase in average global temperature. Those who have contended that human induced warming could be beneficial appear to be concerned with the period before this temperature is reached. In 1938 Callendar argued that, for small increases in global temperature due to fossil fuel consumption there would be the benefits of fuel use, plus the expansion of the northern margin of cultivation, and the delayed return of "deadly glaciers" (Callendar 1938, p. 236). A slight warming of 0.5°C has been estimated to have slight net benefits if air conditioning, agriculture, and water use are considered (about 16 percent of the global economy) (see d'Arge et al. 1975). More recently Idso (1983) has argued that escalated CO_2 concentrations would enhance crop productivity by increasing rates of photosynthesis and water use efficiency by decreasing rates of transpiration; supplying more food for a growing population. Idso has concluded that increased levels of atmospheric CO_2 may actually be beneficial to our future well-being.

During the initial period of warming this qualitative assessment may be accurate. The proportion of climate forcing due to CO_2 is greater than 50 percent compared to other greenhouse gases. Global warming is combined with a CO_2-fertilization effect which may benefit crops and trees. The extent of climate change will be initially within "normal" historical variability. Meanwhile, society currently benefits from the unrestricted and low cost use of fossil fuels and CFCs (ignoring the ozone depletion problem).

As climate change becomes more extreme, however, the benefits are likely to recede and the costs will dominate. The role of non-CO_2 greenhouse gases will become relatively more important over time. As a result the beneficial CO_2-fertilization effect will diminish relative to damaging temperature increases. The more extreme temperature becomes, the greater sea level rise and ecosystem stress can be expected. Yet the absolute temperature change is only part of the problem affecting the intertemporal distribution of costs and benefits. Just as important is the speed of climate change, which will determine how fast impacts escalate and the extent to which forests and unmanaged ecosystems are able to adapt.

There is reason to believe a faster pace of climate change will occur over time because emissions of the principle greenhouse gases are increasing at rates between 0.3 and 5.0 percent per year (Wuebbles *et al.* 1989). As a result, changes in the composition of the atmosphere are taking place in decades, and at an exponential rate (MacDonald 1988). At current emissions rates the earth is being committed to an additional warming of at least 0.15°C and perhaps as much as 0.5°C per decade (Abrahamson 1989 p. 10).

In North America, the fate of numerous tree species will depend on whether they can shift north to cooler climates when their current range becomes uninhabitable. Each 1°C rise in temperature translates into a range shift of about

100 to 150 kilometers (d'Arge *et al.* 1975 and Roberts 1989). The toll would be greater on forests near the southern limits of their current range. The rate of northward dispersal due to historical warming, shown by fossil records, for North American trees is about 10 to 45 kilometers a century. The record for migration is held by Spruce at 200 kilometers a century. Thus, historical scales of forest expansion are much slower than scales of anticipated climate warming. Forest effects should be far more pronounced where climate warming is greater, eg., Canada and Alaska.

The rate of sea level rise also appears set to increase, particularly toward the end of the next century. Titus (1989) has reviewed several studies which analyze sea level rise and Table 3 shows two forecasts for the next century. The marginal rise during the initial twenty-five years is relatively small, regardless the of scenario, compared to the last twenty-five years of the century. This implies a dramatic increase in flooding, salt water intrusion, erosion, and land loss as the century progresses.

The overall picture is one of initial benefits to most regions from slight global warming, but as the climate becomes warmer the likelihood of very large economic costs increases. Population migration will undoubtedly occur as land is lost to rising seas and storm surges, and as semi-arid land becomes unproductive. Wealthier nations with better endowments of fertile soils, less arid and semi-arid land further inland and/or above sea level will be tempted to ignore less fortunate regions (Barbier 1989). The more extreme and rapid the temperature increases the greater are the costs and the fewer are the benefits. Thus, not only will the damages of preceding generations' greenhouse gas releases be placed upon those in the distant future, but the cost of continuing to release those gases will escalate.

Yet, a certain amount of global warming is now irreversible. Concentrations of CO_2 and non-CO_2 greenhouse gases are rapidly increasing in the atmosphere. As a result, within 50 years we are likely to create an irreversible increase in mean global temperature of 1.5°C to 5°C. If no attempt

Table 3: Total Estimated Sea Level Rise in Specific Years (centimeters)

	Future Year				
STUDY	2000	2025	2050	2075	2100
Hoffman *et al.* (1986)					
Low	3.5	10	20	36	57
High	5.5	21	55	191	368
EPA (1983)					
Low	4.8	13	23	38	56
High	17.1	55	117	212	345

Source: Titus (1989) Figure 12.2 p. 169

is made to slow the rate of increase, further 1.5°C to 5°C temperature rise could occur in the following 40 years. Emissions prior to 1985 have already committed the earth to a warming of about 0.9°C to 2.4°C, of which about 0.5°C has been experienced. The warming yet to be experienced is the unrealized warming. This warming, 0.3°C to 1.9°C, is unavoidable (Ciborowski 1989 p. 227-228). The main losers due to the greenhouse effect strongly appear to be the future generations of less developed countries with poor resource endowments.

ECONOMIC MITIGATION STRATEGIES

A 20-30 percent increase in global temperatures, since it has never occurred during recorded history, is outside any practical known limits for sustaining the viability of the present world economy. Since the time frame of change is predicted to be much shorter than historical responses of plants and animals, there is little knowledge as to how ecological systems will adapt or even if they can adapt. The incidence of positive and negative economic effects unlikely to be uniform and may be highly pervasive in that generally, the poorer nations will become even more impoverished and the richer nations relatively richer.[2] Unfortunately, those nations which are contributing the bulk of current CO_2 emissions are also the nations which will benefit from its release in the short to medium term and stand to gain relatively in the long run. This is a kind of worst case situation for managing a global common property resource.

The current polluters benefit in the future from their pollution and the future receptors have little or no current assets to induce polluters to cooperate and will have even less in the future. Thus, no one has the interest or capability to unilaterally attempt to negotiate a solution to this externality problem (see d'Arge & Kneese 1981). The dilemma is one example of the type of problem first posed by Samuelson on over-lapping generations. (Samuelson 1958; see also Gernakoplos and Polmarchakis 1984).

Samuelson suggested the following model structure. The young in each generation receive commodities or income which the old do not. Presumably, the young are productive (and own the assets necessary for producing) while the old are not productive. The old would benefit from income shared by the young and in turn these young would also be compensated by the next generation of young. Unfortunately, there is no mechanism in Samuelson's paradigm for each generation to know that it will be compensated when old, as it had compensated the previous generation when it was young. This is a form of intergenerational negative externality since each generation would be better off by transferring income to the old and in turn receiving such transfers.

The same type of transfer problem arises for the greenhouse problem. The rich nations activities now negatively affect the poor nations in subsequent generations. There is no inherent mechanism for the rich nations to transfer income to the poor now or in the future, even though such compensation might make everyone better off, because, as in the Samuelson case, there is no guarantee that appropriate transfers will be made to the harmed parties in the future.

Figure 1: Paradigm Comparison

Samuelson Paradigm (Y = young; O = old)			CO_2 Paradigm (R = rich; P = poor)		
Y_1	Y_2	Y_3	R_0	R_1	R_2
O_0	O_1	O_2	P_0	P_1	P_2

Without such transfer mechanisms, efficiency in an intertemporal sense is unlikely. The only case where intertemporal efficiency would occur is where each generation accepted as inviolate the distribution of wealth and maximized it's own welfare, regardless of the impacts on other generations. The appropriate transfer mechanism appears to require the existence of an autonomous regulatory agency over succeeding generations and, by definition, continuous governments. This raises the issue where, unlike Samuelson's paradigm which can be resolved within the context of an existing government, there is no such "natural" governmental unit for transfers across governments and time. (see d'Arge and Kneese 1981) That is, the CO_2 paradigm of inefficiency does not lead to a well defined prescription for it's resolution. However, without some form of transfer, the rich nations will become richer at the expense of poor nations in the future. This outcome is only consistent with a "super elitist" ethical view of the global commons where those with the most get the most and those with the least get even less. (see d'Arge, Schulze, & Brookshire 1982)

One of the key problems in the Samuelson overlapping generations model is to provide incentives to an early generation to voluntarily accept the risk and make a transfer regardless of whether a subsequent transfer is made to it (e.g;, they are compensated for its transfer). In the CO_2 paradigm, there is no reward, except in terms of ethics, for the wealthier nations to take this first step. Perhaps this is one of the reasons why we observe a continued call for more research and economic evaluation by the U.S. before taking decisive actions on the greenhouse problem. The only inducement for the wealthier nations, who will be made better off by climatic change, to negotiate might stem from some ethical belief that they should not harm others, without compensation, for their own gain, even though the other nations cannot stop them.

Some economists have argued that negotiation might come about because of interdependent utility or value functions. That is, the wealthier nations citizens are made better off knowing the citizens of impoverished nations are not made worse off by their actions. However, this appears to be somewhat farfetched for impacts to unborn generation of other nations occurring in 100 or so years. The presumption is that we are significantly affected by the negative effects we cause to other nations citizens a long time in the future. With any reasonable individual discount rate for citizens today, the present value of even substantial future damages expressed in individual utility today must be extremely small. There is some preliminary evidence that individuals tend to value impacts on future generations, even their direct heirs, at practically nothing after a century (see Case 1986). If interdependent individual utility functions provide no imperative for preventive actions and/or compensation now for distant future impacts, what can?

In recent years, it has been suggested by Andreoni (1988) and others that individuals when expressing values for public goods, are not deriving their "individualized" Lindahl prices for public goods but rather are expressing a value for concepts of "moral satisfaction". As Kahneman and Knetsch (1992) suggest, individuals are obtaining moral satisfaction as opposed to commodities which elevate their utility through direct or vicarious consumption. By such purchases, individuals make themselves feel good or have a "warm glow" by what they have donated to a good cause. There is substantial empirical evidence that the "warm glow" hypothesis is correct in that individuals are willing to pay a great deal to realize it (see d'Arge 1989). They derive utility not from "consuming" a public good, but in effect, consuming an "ideal". The actual quantities representing the amount of public good purchased or public bad removed is irrelevant. Rather, utility is derived from knowing the right thing is being done. Utility results from the purchase, not from consumption.

The tradeoff between current wealthy individuals and future poor individuals is, thereby, not in terms of reduced wealth to the wealthy and reduced impacts to the future poor, but increased "warm glow" now for less impact on the future poor. Control costs born by the present generation are compensated for through the purchase of moral satisfaction. Sustaining controls through intermediate generations, inclusive of contributions to future compensation, is more likely if these generations also receive "warm glow".[3]

SUMMARY AND CONCLUSIONS

It has been tentatively concluded that two pervasive elements might emerge with regard to climate warming. The first is, on balance, the wealthier nations will benefit and the poorer nations will be harmed. Second, relatively small increases in global temperature will be beneficial globally, but due to the almost irreversible accumulation of CO_2, larger and harmful changes will result. No offsetting actions or compensation for climate warming involve a kind of worst ethical principle being accepted, where the rich become wealthier at the expense of the poor. Such an outcome may be unlikely because of the value citizens place on doing the right thing by purchasing "moral satisfaction". This can be partially accomplished through compensation to future generations through "moral satisfaction" donations now. Further, individual actions on reducing use of fossil fuels, directly or indirectly, can contribute to reducing intertemporal inefficiency. Such individual actions are not enough, but at least they would be a step in the right direction.

NOTES

1.   This outline of regional impacts is partially derived from the conclusions of the workshops held in Villach Austria (28 September - 2 October, 1987), and Bellagio, Italy (9-14 November, 1987), under the auspices of the Beijer Institute, Stockholm (World Meteorological Organization 1988, See Jaeger 1989 for a summary).

2.   The greenhouse effect is characterized here as leading to gains for the rich nations while the poor lose. In fact, as argues earlier there are likely to be serious impacts on all nations in the long run, but the rich nations remain relatively better off. That is, as the cake shrinks the rich nations end up with a larger percentage. For simplicity in the following discussion and model we assume the rich gain absolutely, even in the long run.

3.   In order to place these ideas in a more formal structure, a rather simple model is outlined in Appendix A. The model structure embodies, in elemental form, the concepts just outlined in a rather simplistic way.

REFERENCES

Abrahamson, D.E., "Global Warming: The Issue, Impacts, Responses," in The Challenge of Global Warming edited by D.E. Abrahamson (Washington, D.C.: Island Press, 1989): 3-34.

Adams, R.M., B.A. McCarl, D.J. Dudek, and J.D. Glyer, "Implications of Global Climate Change for Western Agriculture," Western Journal of Agricultural Economics 13, no.2 (December 1988):348-356.

Andreoni, J., "Giving with Impur Altruism: Application to Charity and Ricardian Equivalence," Journal of Political Economy, Vol. 97, No.6 (1989).

Barbier, E.B., "The Global Greenhouse Effect: Economic Impacts and Policy Considerations," Natural Resources Forum (February 1989):20-32.

Broadus J.M., J.D. Milliman, S.F. Edwards, D.G. Aubrey, and F. Gable, "Rising Sea Level and Damming of Rivers: Possible Effects in Egypt and Bangladesh," in Titus, Vol. IV, 1986.

Callendar, G.S., "The Artificial Production of Carbon Dioxide," Quarterly Journal of the Royal Meteorological Society 64 (1938): 223-240.

Case, J.C., Contributions to the Economics of Time Preference, doctoral dissertation, (Department of Economics, University of Wyoming December 1986).

Ciborowski, P., "Sources, Sinks, Trends, and Opportunities," in The Challenge of Global Warming, edited by D.E. Abrahamson (Washington D.C.: Island Press, 1989): 213-230.

d'Arge, R.C., editor, Economic and Social Measures of Biologic and Climatic Change, Climate Impact Assessment Program, (prepared for the U.S. Department of Transportation, Washington, D.C. 1975).

d'Arge, R.C., W.D. Schulze, and D.S. Brookshire, "Carbon Dioxide and Intergenerational Choice," American Economic Review 72, no.2 (May 1982): 251-156.

d'Arge, R.C., "A Practical Guide To Economic Valuation of the Natural Environment," presented at the Rocky Mountain Mineral Law Foundation Conference, Snowmass, Colorado, July 1989, to be published in The Rocky Mountain Mineral Law Institute, forthcoming 1990.

d'Arge, R.C., and A.V. Kneese, "State Liability for International Environmental Degradation: An Economic Perspective," Natural Resources Journal Vol. 20, no.2 (July 1980): 427-450.

Dregne, H.E., Desertification of Arid Lands (Chur, Switzerland: Hardwood Academic Publishers, 1983).

Gates, D.M., "An Overview," in CO₂ and Plants: The Response of Plants to Rising Levels of Atmospheric Carbon Dioxide, edited by E.R. Lemon, (Boulder, CO: Westview Press, 1983): 7-20.

Geanakoplos, J., and H.M. Polmarchakis, "Intertemporally separable overlapping generation economics," Journal of Economic Theory 34 (1984).

Glantz, M.H., "Assessing the Impacts of Climate: The Issue of Winners and Losers in a Global Climate Change Context," in Changing Climate and the Coast, J. Tutus, ed. (Washington, D.C.: U.S. EPA, 1990) forthcoming.

Goemans, T., "The Sea Also Rises" The Ongoing Dialogue of the Dutch with the Sea," in Titus, Vol. IV, 1986.

Grotch, S.L., Regional Intercomparison of General Circulation Model Predictions and Historical Climate Data, (Washington D.C.: Department of Energy, 1988).

Hansen, J., A. Lacis, D. Rind, G. Russell, I. Furg, P. Ashcroft, S. Lebedeff, R. Ruedy, and P. Stone, "The Greenhouse Effect: Projections of Global Climate Change," in Effects of Changes in Stratospheric Ozone and Global Climate, edited by J.G. Titus (U.S. Environmental Protection Agency, Vol. I, August 1986).

Heathcote, R.L., The Arid Lands: Their Use and Abuse, (Harlow, England: Longman, 1983).

Idso, S.B., "Carbon Dioxide and Global Temperature: What the Data Show," Journal of Environmental Quality 12, no. 2 (April-June 1983): 159-163.

Jaeger, J., "Developing Policies for Responding to Climate Change," in The Challenge of Global Warming, edited by D.E. Abramson (Washington D.C.: Island Press, 1989).

Kahneman, D. and J. Knetsch, "Valuing Public Goods: The Purchase of Moral Satisfaction", Journal of Environmental Economics and Management, v22(1), pp 57-70.

Kelejian, H.K. and B.V. Varichek, "Pollution, Climate Change, and Consequent Economic Costs Concerning Agricultural Production," in Economics of Managing Chloroflurocarbons: Statospheric Ozone and Climate Issue, edited by J.H. Cumberland, J.R. Hibbs, and I. Hoch. (Baltimore: RFF, John Hopkins Press, 1982).

Kneese, A.V., "Confronting Future Environmental Challenges," in Resources 99 (Spring 1990).

Kosobud, R.F., and T.A. Daly, "Global Conflict or Cooperation over the CO_2 Climate Impact." Kyklos 37 (1984):638-659.

MacDonald, G.J., "Scientific Basis for the Greenhouse Effect," Journal of Analysis and Management 7, no.3 (1988): 424-444.

"Maldives: A Sinking Feeling," The Economist (1st October, 1988): 36

Newman, J.E., and R.C. Picket, "World Climate and Food Supply Variations," Science, 186, no.4167 (1974): 877-881.

Oguirtoyirbo, J.S., and R.S. Odirgo, "Climatic Variability and Land Use: An African Perspective," (presented at World Meteorological Organization, World Climate Conference, Geneva, 12 to 23 February, 1979).

Park, R.A., T.V. Arentaro, and C.L. Cloonan, "Predicting the Effects of Sea Level Rise on Coastal Wetlands," in Titus, Vol. IV, 1986.

Parry, M.L. and T.R. Carter, "Effects of Climatic Changes on Agriculture and Forestry: An Overview," in Titus Vol. I, 1986.

Roberts, L., "How Fast Can Trees Migrate?" Science 243 (February 1989): 735-737.

Rosenzweig, C., "Potential CO_2 Induced Climate Effects on North American Wheat Producing Regions," Climate Change 7 (1985):367-389.

Samuelson, P.A., "An Exact Consumption Loan Model of Interest, With or Without the Social Contrivance of Money," Journal of Political Economy 66, no.6 (1958).

Smagorinsky, J., "Effects of Carbon Dioxide," in Changing Climate (National Research Council, Washington, D.C.: National Academy Press, 1983).

Smit, B., L. Ludlow, and M. Brklacich, "Implications of a Global Warming for Agriculture: A Review and Appraisal," Journal of Environmental Quality 17, no.4 (1988) 519-527.

Spash, C.L., and R.C. d'Arge, "Compensation of Future Generations for Adverse Climatic Changes", submitted to Georgetown International Environmental Law Review, April 1990.

Thomas, R.H., "Future Sea Level Rise and its Early Detection by Satellite Remote Sensing," in Titus, Vol. IV, 1986.

Titus, J.G., editor, Effects of Changes in Stratospheric Ozone and Global Climate, (U.S. Environmental Protection Agency, August 1986).

Titus, J.G., "The Cause and Effects of Sea Level Rise," in The Challenge of Global Warming, edited by D.E. Abrahamson (Washington D.C.: Island Press, 1989): 161-195.

Wallen, C.C., "Arid Zone Metrology," in Arid Lands: A Geographical Appraisal, edited by E.S. Hills (London: Methuen and Co. Ltd., 1966): 53-76.

White, G.F., "The World's Arid Areas," in Arid Lands: A Geographical Appraisal, edited by E.S. Hills (London: Methuen and Co. Ltd., 1966): 19-30.

World Meteorological Organization, World Climate Program, Developing Policies for Responding to Climatic Change, A Summary of the Discussions and Recommendations of the Workshops held in Villach, Austria (28 September - 2 October 1987), and Bellagio, Italy (9-13 November 1987), edited by J. Jaeger, (World Meteorological Organization, WCIP-1, 1988).

Wuebbles, D.J., N.E. Grant, P.S. Connell, and J.E. Penners, "The Role of Atmospheric Chemistry in Climate Change," Journal of Air Pollution Control Association 39, no. 1 (January 1989): 22-28.

APPENDIX - THE MODEL

Let there be two time periods, 1 and 2, where 1 denotes the present and 2 the future. Assume there are two countries, R (Rich) and P (Poor), each with an endowment of resources which determines consumption in the two time periods. The wealth of each countries citizens can be represented by a single, quasi-concave utility function u(·). The normal argument is consumption in each country in each time period e.g., C^{R1}. The first order conditions for an optimum can be simplified to

$$\frac{u_c^{R1}}{u_c^{R2}} = N^R(1+x^R)$$

where N^R is the utility discount factor for country R from period 1 to period 2, and X^R is the productivity on investment earned from period 1 to 2 per unit of consumption foregone in period 1. Country P has an identical set of conditions as long as there is no externality between the two countries. This condition is commonplace, in that all that is established is: the marginal rate of substitution between present and future consumption equals the utility discounted productivity of foregone consumption.

Next introduce an external diseconomy, where C^{R1} negatively influences the levels of C^{P2}. If the two countries act independently then country R will continue to use first order condition (1) to maximize utility, ignoring the effect on country P. The first order conditions (1) under social optimality change because of the presence of the externality; the independent choices implied in (1) are no longer optimal. Joint maximization of utility across countries makes the optimal condition for country R:

$$\frac{u_c^{R1}}{u_c^{R2}} = N^R[(1+x^R) + \frac{\theta F_c^{R1}}{\delta^{R2}}]$$

An additional term has been now added to the RHS. This term includes δ^{R2}, a Lagrange multiplier for consumption in the future period, θ a multiplier on the externality and F_{CR1}, the reduction in C^{P2} resulting from a unit increase in C^{R1}. The addition of this term is a standard result in externality theory. The usual procedure is to levy a tax on C^{R1} equivalent to this negative effect and thereby bring about a change of first order conditions from (1) to (2).

The first order condition for the P country is:

$$\frac{u_c^{P1}}{u_c^{P2}} = N^P[(1+x^P)\frac{\delta^{P2}}{\delta^{P2}-\theta}]$$

Note, a new term involving multipliers also appears in the first order conditions for the P country as well as the R country. For an optimum to be achieved, joint maximization is required over both countries, and adjustments in first order conditions of both countries have occurred.

Now, let us add the further complication that not only does increased C^{R1} reduce C^{P2}, but also raises C^{R2}. Consuming fossil fuels at higher rates now not only reduces possible consumption of the poorer nations in the future, but raises income-consumption of the wealthier nations in the future. Thus, there is an intertemporal diseconomy coupled with an intertemporal positive economy. The first order conditions of country R acting independently become:

$$\frac{u_c^{R1}}{u_c^{R2}} = N^R(1+x^R-g_c^{R1})$$

and the first order social optimum condition for country R. is:

$$\frac{u_c^{R1}}{u_c^{R2}} = N^R(1+x^R-g_c^{R1}+\frac{\theta F_c^{R1}}{\delta^{R2}})$$

where g_c^{R1} is the change in C^{R2} resulting from an increase in C^{R1}. Note that current investment through reduced current consumption now has a penalty attached to it reflecting the loss of wealth by not causing greater climatic change.

The important point in this simple model is that efficiency is impossible the two county context unless a) a Pigouvian tax is levied from outside, or b) country R voluntarily agrees to impose the tax on itself, even though this would reduce its future wealth from climatic change. Neither of these possibilities seems very realistic since there is no international authority with the power to enforce the tax, and no apparent economic motive for the R country to voluntarily comply.

Now let us introduce the idea of Andreoni-Kahneman-Knetsch, that individuals desire to do the right thing and are willing to purchase "moral satisfaction". The utility function for the R country now includes an argument for not depressing consumption in the P country. Such an argument can be modeled in a number of ways, but the general idea is that citizens of the R country will make contributions to the P country to offset P country losses from R country activities, and R country citizens receive utility from this "act" of giving.

The effect of this is to create a new value in the optimum which partially or completely offsets the negative impacts on country P, and provides country R with a motivation not to indiscriminately harm country P. The first order conditions for country R acting independently become:

$$\frac{u_c^{R1}}{u_c^{R2}} = N^R(1+x^R) - \frac{u_{cP2}^{R1} \frac{dc^{P2}}{dc^{R1}}}{\delta^{R2}}$$

which is structurally similar to (2). Since dc^{P2} / dc^{R1} equals f_2^{R1}, we see the (6) and (2) are identical if U_{cp2}^{R1} equals x in the joint maximization. This is unlikely to occur except by pure chance.

The introduction of a moral motivation induces country R to independently take actions that bring us closer to a social optimum. The 1990 Earth Day theme stressing individual actions and responsibilities is one example of the potential desire to purchase moral satisfaction.

CHAPTER
V

THE EFFECTS OF GLOBAL WARMING ON THE MINING INDUSTRY: ISSUES, TRADEOFFS & OPTIONS

H. Stuart Burness
&
Wade E. Martin

INTRODUCTION

Global warming has become one of the primary topics of environmental concern in recent years. The source of this concern arises from confusion and uncertainty over exactly what global warming entails, where and to what degree it will occur, what the implications of global warming are in terms of physical effect, climate change, weather patterns, and the resultant economic and environmental changes. Predictions and forecasts of these physical effects of global warming are certainly not lacking. However, in spite of the apparent consensus of the general circulation models, there are still many who question the results of the models on the basis of incomplete scientific information and analysis. We do not intend to enter this debate, but, rather wish to draw inferences from it concerning the actual and/or possible economic implications of global warming, particularly for the mining sector.

The concerns of the mining sector regarding potential global warming are often implicit if not explicit. These concerns arise in part due to an imprecise understanding of the implications of global warming by the mining community. Anticipating what is to come, it is important to note the distinction between the causes and effects of global warming. In this regard it appears that warming. However, the fossil fuels sector of the mining industry is involved as a suspected cause of global warming through the release of carbon dioxide (CO_2). Consequently, the effects on the mining industry may be felt indirectly through policies aimed at limiting CO_2 release. The purpose of this investigation is to ascertain the general nature of global warming and the implications for the mining industry. In order to do this we will explore the physical aspects as well as the economic aspects of global warming. Some tentative conclusions will be presented concerning the likely future of the mining industry as affected by various policies that address global warming or the "greenhouse effect."

The number of economic analyses has been relatively sparse compared to the volumes of scientific studies on global warming, or the greenhouse effect. Until recently, economists who have been contemplating this issue seem to have taken a somewhat unified stance concerning the potentially devastating impacts of the long run effects. This unified stance contrasts with Nordhaus' (1977) view that "(natural) scientists are divided between one group crying 'wolf' and another which denies that species' existence." However, more recent economic studies suggest a shift in the economists' outlook to reflect some of the diversity of opinion exhibited by the physical scientists.

The uncertainty facing the environmental scientist is compounded due to the limitations involved with modelling the environment and climate change. "Experiments" must be undertaken in the context of computer modelling and simulation as real world experiments are impossible. Modelling and simulation procedures employed in the environmental sciences are generally quite sophisticated, but in some areas, such as the effects of urbanization, cloud and ocean temperature diffusion models, the surface is just being scratched.

Realizing the limits of the climatological models is critical to evaluating economic impacts of the potential climate change. Therefore, Section 2 will address the physical aspects of the global warming issue. The climatological modelling process and the uncertainty involved in such modelling will be discussed. Section 3 will then develop the economic modelling appropriate to analyze the common property nature of such a pollution problem. This will be followed by a discussion of the potential impacts of global warming on the mining sector of the economy. A summary and concluding section will then review the main results.

THE PHYSICAL NATURE OF GLOBAL WARMING

The mean temperature of the Earth has been constantly changing. The concern now is whether the temperature change is a naturally occurring event or a change induced by human activity. The physical side of the problem is to measure and estimate the change in mean global temperature, the regional nature of the change and the origin of this change.

To understand global warming, it is necessary to consider the transfer of heat between outer space, the Earth's atmosphere and the Earth's surface. The Earth's mean temperature is estimated to be -18 degrees Celsius in the absence of an atmosphere. However, the presence of an atmosphere generates a 'greenhouse' effect that warms the Earth to a mean global temperature of 15 degrees Celsius. The greenhouse effect is generated by the presence of the so-called 'greenhouse gases' (GHG).

The greenhouse gases are important because they are inefficient absorbers of incoming solar or short-wave radiation yet are quite efficient at absorbing the outgoing long-wave radiation emitted from the Earth's surface. The degree to which the long-wave radiation emitted from the Earth's surface is trapped in the atmosphere results in the warming of not only the atmosphere but also the Earth's surface.

There are numerous greenhouse gases, however, the most important for global warming are water vapor, carbon dioxide, methane, nitrous oxide and chloroflurocarbons.[1] The presence of each of these gases in the atmosphere has been measured or estimated using 1860 as the base year. This year was used to coincide with the industrial revolution in the United States since these gases, excluding water vapor, are highly correlated with industrialization. Table 1

Table 1: Greenhouse Gas Concentration (ppm)

GHG	1860	1986	2035
CO_2	290	348	475
Methane	1.1	1.7	2.8
Nitrous Oxide	0.28	0.34	0.38
CFC's	0.00	0.006	0.005

presents the concentrations of these gases for the years 1860, 1986 and estimates for 2035.

The contribution of each of these gases to global warming varies. Carbon dioxide is by far the major contributor to the warming trend, both from an historical perspective as well as for future projections. The contribution of the other gases is not consistent over time. When considering the increase in mean global temperature between 1860 and 1986 methane and nitrous oxide account for the balance of the warming. Whereas, the projections to 2035 show a dramatic increase in the contribution of the CFC gases, particularly CFC12. This is due to the efficiency of CFC's in trapping long-wave radiation emitted from the Earth's surface.[2]

Since many GHG's are highly related to the level of industrialization it is useful to consider the causes of increased concentrations of these gases. In this context, the primary gases of concern are carbon dioxide and CFC's. The focus here is on carbon dioxide, which is consistent with the climatological models that base their projections on increased concentrations of CO_2.[3]

The dramatic increase in the atmospheric concentrations of CO_2 has been attributed to two primary sources, deforestation and the burning of fossil fuels. Deforestation has reduced an effective means of removing CO_2 from the atmosphere. Trees use CO_2 and then release oxygen to the atmosphere thus decreasing the concentration of CO_2 in the atmosphere. Reforestation as a policy option is the main approach to address the problem after the fact. This will be considered in more detail below. Currently, the main sink for CO_2 in the environment are the oceans. However, the rate of uptake is not sufficient to account for all the release of CO_2 to the atmosphere.

The concentration of CO_2 in the atmosphere closely parallels the release of CO_2 from the burning of fossil fuels. To reduce global warming by a significant amount requires controlling the concentration of CO_2 in the atmosphere. To accomplish this, it is necessary to either increase the sinks for CO_2 (e.g., more forests) or to reduce the emissions of CO_2 (e.g., reduce the burning of fossil fuels). Section 3 addresses the economic aspects of such policies.

To this point we have focused on the increase of the mean global temperature. The variation of the temperature changes also of importance. This involves two aspects of temperature change, distributional changes in mean temperature by geographic region and changes in the deviations of these mean temperatures. The climatic models used to estimate global warming do not all agree on all aspects of the geographic effects of global warming.

Considering the geographic effects first, impacts on the mean global temperature change given a doubling of the present CO_2 emissions are estimated. Although the models do not agree regarding the change in mean temperatures, they are more consistent in identifying the areas that will be affected most severely. The area experiencing the most severe temperature (not necessarily economic) change is the northern latitudes, approximately 70 degrees N.[4] The geographical changes will have a significant impact on the distribution of economic activity. This issue will be considered in Section 3.

The second issue that the climate models address is the standard deviation of the mean global and regional temperatures. (Mitchell, 1989; Schneider and Rosenberg, 1988) It is argued that along with the increase in the mean temperatures, there will be an associated increase in the deviation of these temperatures. One implication of this, for example, involves the occurrence of a drought. The duration of the drought may be greater than is currently the case thus increasing the severity, both economically and physically.

Although the physical aspects of global warming are known,[5] the predictive ability of climatological models are much less reliable. The ability of the models to include all the relevant parameters, such as the effect of oceans, clouds and urbanization on global warming, has lead to serious doubts about the future projections of these models (Bryson, 1989). At this time, however, the general consensus of climatologists is that global warming is indeed occurring but the debate concerns by how much and whether or not it is part of the natural climate cycle or human-induced. Given that warming is apparently occurring and that the greenhouse gases are the cause, it is natural to inquire as to the economic consequences and the policy options for limiting global warming and/or mitigating its effects.

A HEURISTIC ECONOMIC MODEL OF GLOBAL WARMING

An economic analysis of global climate change is both an extension and an adaptation of the atmospheric-environmental or climatological model. Figure 1 depicts an overview of the economic aspects of global warming relative to underlying atmospheric models. Changes in emissions sources and sinks affect atmospheric changes which through the atmospheric model result in a change in climate. This then translates into geophysical environmental effects which imply in part a sequence of economic effects. Economic effects are broken down into direct and indirect effects. The direct effects involve gainers and losers who are easily identified and benefits and costs which are estimated using market values. Indirect effects involve recipients who may not be easily identified, except perhaps as a group. These indirect effects also include public or non-market benefits and costs for which there are generally no market data available.[6]

In terms of received economic theory the predominant characteristic of an economic analysis of global warming is the presence of a common property resource...in spades. The common property resource of concern is the atmosphere itself. There are a number of classic cases of common property resources. A brief inquiry into the nature in which they have been resolved or at least attempted to be resolved is illuminating in the context of the problem at hand. Generally, the common property resource is one which is not

Figure 1: The Interaction of Economic Effects and Abatement Policy of the Greenhouse Effect

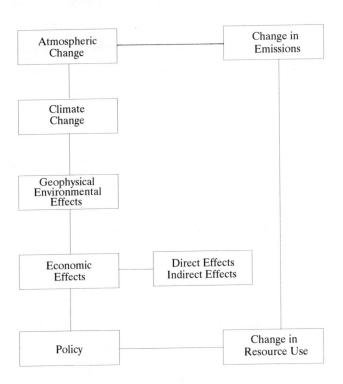

appropriated, or may even be unappropriable, and is usually compounded by the presence of stock effects. Thus, since the resource is "free" each individual effectively ignores the effect of his or her use of the resource in terms of its effect on the stock. But summed over all individuals these stock effects can be substantial and result in the depletion of the resource. Using the basic common property models as a foundation, an economic model of global warming will be presented. This model will then provide the necessary background for analyzing the impacts of global warming on the mining sector.

Perhaps the most revealing economic approach to this problem is that undertaken by Nordhaus (1989). The problem with discounting notwithstanding, Nordhaus attempts to incorporate the available scientific evidence into a representative benefit-cost analysis in order to gain insights regarding the optimal control strategy. This, of course, is done with the realization that forecasting future climate changes is highly uncertain. This uncertainty is due to the long forecasting period for emissions of greenhouse gases and the lagged and ultimate effects of these emissions on the climate, as discussed above.

The realization that the greenhouse problem, when viewed as a common property problem, is further complicated to the extent that emitters of atmospheric gases impose costs on others and that these costs are not ultimately borne until some time in the distant future. These costs would be appropriately measured by what Nordhaus calls the "greenhouse damage function." This damage function incorporates numerous economic impacts such as changing crop yields, migration of population, altered recreational patterns, various environmental costs, and the costs associated with rising ocean levels.

Given earlier discussions of the uncertainties concerning the nature, degree, timing and bounds on the greenhouse effect, there are certainly analogous types of uncertainties associated with the actual economic effects. Moreover, the regional variations in greenhouse effects raise questions as to whether net benefits may in fact be positive. Understandably, Nordhaus is most reluctant to speculate on the nature of the greenhouse damage function, but, as a last and only resort, presumes that it has the usual classical form. In fact at this juncture we will switch terminology and refer to this as the greenhouse benefit function; that is, the relationship that describes the benefits received from reduced CO_2. In contrast to the uncertainty surrounding the greenhouse benefit function, Nordhaus suggests that much more is known about the abatement cost function, which reflects the costs that society must incur to mitigate or adapt to the greenhouse affect. These would include such costs as the cost of finding substitutes for fossil fuel, finding substitutes for chlorofluorocarbons, costs of reforestation programs, and other programs intended to mitigate GHG's. Thus the abatement cost function considers only the costs of such programs that are intended to reduce GHGs and/or effects of global warming. Costs such as those associated with building dikes to prevent sea water encroachment, increased air conditioning costs, and increased irrigation costs are properly included in the greenhouse benefit function as costs avoided.

This bifurcation allows a clear distinction between the greenhouse benefit function and the abatement cost function. The items fraught with uncertainty, measurement problems, and problems with regional variations are included in the greenhouse benefit function, while those items for which there is a clearer analysis of effects and a more quantifiable aspect to costs are included in the abatement cost function.

The optimal level of GHG concentrations given the status quo level of concentrations SQ_0 is at Q_0 as determined by the intersection of marginal greenhouse benefits (MGB) and the marginal abatement cost (MAC_0) at E_0 in Figure 2. Net social benefits of GHG reductions are given by the area between MGB and MAC_0 as bounded on the left by SQ_0, in this case area A. At a higher level of initial concentrations such as SQ_1 the schedule MAC_1 becomes relevant in terms of expressing the marginal abatement costs, and the optimal level of GHG concentrations is between Q_L and Q_H as determined by E_L and E_H, respectively. In this case, net social benefits from the reduction of CO_2 have a lower bound as given by area B and an upper bound as given by areas B + C.

The CO_2 concentrations measured on the horizontal axis of Figure 2 are equilibrium steady state levels. That is, absent any policy changes with respect to CO_2, SQ_0 is not necessarily the current level of concentrations, but the steady state level of concentrations given the current level of emissions. In addition, the abatement cost function is defined relative to steady state concentrations and it is not clear what relationship MAC_1 would bear to MAC_0.

Even more uncertainty is involved concerning the MGB schedule in terms of its height and slope at higher levels of concentrations. This is due to questions concerning the actual extent and nature of potential damages from increased concentrations as well as the many lagged effects in the atmospheric and climatological models of CO_2 concentrations. As an example consider

Figure 3. The left panel reflects two different levels of emissions: E_1 the current level of emissions, and E^* the "optimal" level of emissions. This is an oversimplification in the sense that the optimal level of emissions may not be constant over time, however, it is still a useful heuristic. Associated with E_1 is a steady state level of CO_2 concentrations as shown by the dotted line at S_1 in the right panel. The path P_1 shows a possible approach to S_1. Likewise, associated with E^* is an optimal steady state level of concentrations shown by S^*. P^* represents a possible approach to S^* for a given policy adopted at the current time t_0. However, if the level of emissions E^* were not affected until time t_1, the path to S^* may be quite different. One would expect that this path would be a deviation from P_1, but might be represented by a path such as P^{*B} which is asymptotic to S^* from below, or a path such as P^{*A} which is asymptotic to S^* from above. Of course there are other possibilities as well, but the point to be made is that even though the steady state level of concentrations is the same in either case, damages might be quite different depending upon which path to S^* was relevant.

In spite of the above difficulties, this simple model yields some interesting insights into the nature of the problem. Also, regardless of the many uncertainties concerning the benefit function, Nordhaus is able to make some interesting observations concerning policy options in terms of the abatement cost function. Nordhaus considers four possible policies for reducing GHG emissions: i) reduce CFC emissions, ii) reduce CO_2 emissions, iii) reforestation, and iv) gasoline taxation. He concludes that these various policies are quite different in their relative cost-effectiveness in reducing GHG emissions. The most cost-effective option is reducing CFC emissions. CFCs are many orders of magnitude more efficient in absorbing long wave radiation than CO_2, and there are relatively low cost substitutes for a number of the CFCs. As a result, only a modest curbing of CFC emissions are necessary to obtain significant benefits in the form of reduced global warming. This approach is currently being instituted as a result of addressing the problem

Figure 2: Benefits and Costs of Abatement

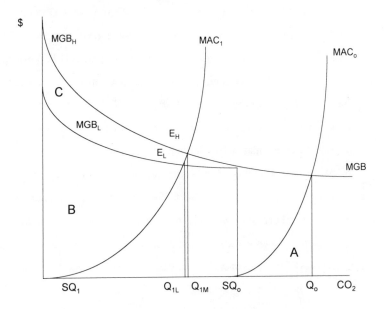

Figure 3: Emissions/Concentrations Time Paths

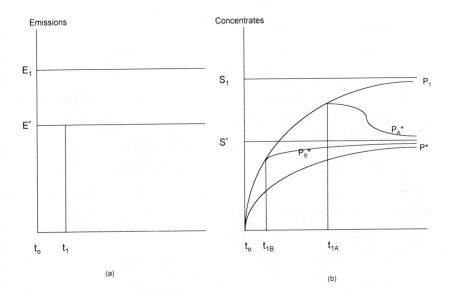

associated with depletion of stratospheric ozone.

The next most efficient option is the reduction of CO_2 emissions through the use of a carbon tax. This is a tax on the carbon content of fossil fuels. A gasoline tax is another form of the carbon tax which is less effective and applies only to automotive sources and hence, has only a fraction of the effect. The reforestation option is the least effective of all, as the marginal cost per ton of carbon fixed is extremely high. The marginal cost of these different abatement policies as estimated by Nordhaus is depicted in Figure 4.

Nordhaus also attempts a similar exposition concerning the composition of the greenhouse benefit function. He identifies agriculture and forestry as two of the sectors most likely to be affected by climate change and includes EPA (1988) assessments of such effects. He also includes costs associated with sea level rise, increased energy demands, for a variety of market goods and services as well as various non-market goods and services. While the costs of these last two categories are considered, they are not specifically included due to uncertainties, regional variation or absence of market values. On the basis of the high level of uncertainty concerning the benefits' side of the analysis, Nordhaus summarizes the results in scenario form with a marginal benefit of $3 per ton CO_2 equivalent reduction in the middle case and a marginal benefit of $37 per ton CO_2 equivalent reduction in the high case.

Combining benefits and costs with abatement options coming on line in terms of their relative efficiency, the optimal solution to the medium case scenario occurs at approximately a 20% reduction in emissions from the uncontrolled level, and the high case represents a reduction of about 31%. The reduction in emissions for the middle case are almost entirely through reduction in CFC emissions with about a 10% reduction in CO_2 emissions. The high case scenario requires that CFC emissions are completely eliminated and CO_2 emissions are reduced by about 25%. In both cases the reforestation option is not economically efficient.

Nordhaus is quick to point out several shortcomings in this analysis.

Figure 4: Marginal Cost of Emissions Reductions by Policy Type

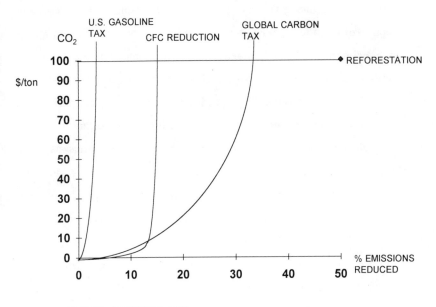

SOURCE: NORDHAUS, 1989

First, the abstraction from economic and climactic interactions by viewing the process as a single "snapshot" of the world. Second, the assumption that the same sectoral composition of the economy will exist in the future as it does now. Third, this analysis ignores the effects of economic discounting. Finally, the analysis also ignores other sources of market failure such as ozone depletion or air pollution. Nonetheless, the analysis forms a useful basis for evaluating some of the policy aspects regarding the effects of global warming on the mining sector.

EFFECTS ON THE MINING INDUSTRY

It appears safe to say that the <u>direct</u> aggregate effects on the mining sector as a result of global warming would be minimal. However, there is a strong link between mining, primarily in the energy (fossil fuel) sub-industries, and global warming due to the release of CO_2 in the mining, production, and consumption of fossil fuels. The effect on the mining industry will be indirect in that policies intended to limit emission of CO_2 may impose costs and limitations on the consumption of fossil fuels and perhaps even on the production of fossil fuels. These costs will be passed back to the energy sector and, to the extent that these industries are competitive, these effects will ultimately be the same regardless of their policy origin. Also, global warming policies may have distributional effects due to substitution within the mining sector (e.g., uranium mining v. coal mining) and by use (e.g., industrial minerals for wood products).

The indirect impact on the mining sector depends not only on the policy options pursued, but also on which marginal benefits scenario proves correct. If the $3 per ton of CO_2 (equivalent) emissions proves correct, then even the indirect effects on the mining sector would be minimal. However, if the high scenario of $37 per ton results then the indirect effects could be significant for the mining industry.

A number of studies purport to model the effect of emissions on atmospheric CO_2 concentrations based on economic criteria that focus on the role of economic parameters in terms of their effect on demand and supply of various energy types and technologies, and their ultimate effect on CO_2 emissions. Edwards and Reilly (1989) present one of the more recent studies which provides an interesting reference point with regard to these types of analyses.

Their focus is on the "doubling date" of CO_2 concentrations which occurs at concentrations of 600 parts per million (ppm). There is nothing magical in terms of a "doubling" of concentrations as opposed to other values, this is just picked as a convenient reference point. In reality, some "greenhouse" effects may occur prior to the doubling of concentrations and others perhaps after doubling occurs.

The basic idea of the analysis is to focus more closely on the interaction between energy, economic and demographic factors with less complication in terms of the climatological aspects of modelling.   This focus allows the determination of the effect of energy prices, supply and demand, and energy balance in terms of what fuels will be used, what the related carbon release will be, and how energy balance will be achieved. This information is then used to determine the time path of emissions from which one can infer the level of CO_2 concentrations at any point in time.

Edwards and Reilly (1989) predict a doubling of CO_2 concentrations in roughly 60-80 years, depending upon the scenario. They consider the CO_2 coefficients of various fuels, demand income, price and cross elasticities, population and GNP growth rates for various regions (global), supply (resource) restrictions for constrained technologies, and breakthrough costs for unconstrained technologies. Their results indicate relatively stable electricity prices due to substantial increases in coal-fired and nuclear power. This implies that the future for coal and uranium mining would be improved as well. However, this must be tempered with their estimate that the U.S. share of global energy falls from a current 29% to 13% of global use by 2010.[7]

Developing areas energy consumption is predicted to increase from a current 19% to 52% of global power consumption by 2010. Overall, primary energy use falls from a current 4.5% to 2.5% per year. The release of CO_2 increases, however, from 1.5% in 1975-2000 (only one-third of the historical average) to 2.3% in 2000-2025 to 3.1% in 2025-2050. These changes are determined simultaneously by changes in energy consumption, shifting in energy production patterns, and by the introduction of high CO_2 oil obtained from carbonate rock in the 2025-2050 period. In effect, even though less energy is derived from fossil fuels, the average fossil fuel becomes dirtier over time, especially beyond 2025.

A comparison of carbon emissions among four major energy studies is presented in Figure 5. These studies were performed by Nordhaus (1977), Hafele (1981), Rotty and Marland (1980), and the above cited study by Edwards

Figure 5: Alternative Projections of Carbon Emissions

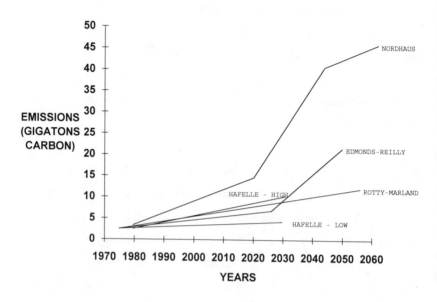

Source: Edmonds & Reilly, 1989

and Reilly. The notable similarities and differences in these studies are i) the similarities in projections, except for base case assumptions, between 1975-2025; ii) the deviations in projections from 2025 on, with Edwards-Reilly and Nordhaus showing increased carbon emissions and Rotty-Marland and Hafele showing a decrease in the rate of emissions increase.

The next step after determining carbon emissions is to infer CO_2 concentrations. Edwards and Reilly show CO_2 concentrations increasing from a rate of roughly .2% per year in 1900-1975 to .38% in 1975-2000, .55% in 2000-2025, finally to .91% in 2025-2050.[8] On the basis of these calculations, they determine the doubling date of CO_2 to be between 2045 and 2072. Their primary observation is that movements in the doubling date are relatively insensitive to changes in various parameters... the magic of exponential growth.

An interesting part of their analysis is the construction of alternative scenarios concerning various policy responses to increased CO_2 emissions and concentrations. The first scenario imposes taxes of 50% on coal, 39% on oil, and 28% on gas, with the taxes being applied to end use. While this has a noticeable effect on U.S. carbon emissions, it has almost no effect on global CO_2 emissions as the reduced U.S. demand for energy leads to a lowering of global energy prices and hence increased use. The delay in the doubling time can be measured in weeks!

Scenario 2 differs in that the taxes are applied to carbon release and U.S. exportation of coal is prohibited. While there are more noticeable changes in both U.S. and global carbon emissions, the effect on the doubling date is to delay it only by about five years. Finally, in scenario 3 the relevant tax rates are applied globally. In this case carbon emissions are reduced by 40% in 2050 as compared to the base case scenario. This results, however, in delay of the doubling by less than 10 years. In all cases the effect on U.S. GNP is minimal, less than a .5% reduction.

While scenario 3 is largely academic in terms of the likelihood of achieving it, the analysis still points to the fact that even massive policy changes may have only negligible effects in terms of the doubling date. This points toward benign energy technologies as the more likely means to provide a shift away from the base case scenario.

The implication for the mining industry, or at least its energy component, seems to be fairly straightforward. The energy (or CO_2) explosion predicted by these analyses, if it occurs, will begin around the second quarter of the 21st century. At that time the economics of energy would prefer reliance on coal and oil shale, but other considerations, as discussed above, would suggest more benign technologies. The interim, in a sense, could be viewed as a period of grace, from that point on the future is more uncertain. It would seem unwise at this juncture to suggest exactly what those uncertainties involve, but the concerns are real, and as time passes, efforts should be directed towards assessing these issues.

SUMMARY AND CONCLUSIONS

An overview of the so-called "greenhouse effect" or global warming has been presented, first in terms of what is known about this phenomenon from the physical sciences, next in terms of the economic aspects or these potential and possible changes, and finally in terms of the implications for the mining industry. As a consequence, it becomes increasingly clear that the keynote of these endeavors is uncertainty. Original prophecies of doom are more recently being replaced by the suggestion that as we learn more about this phenomenon we find that there is much more to know. However, there is no consensus on this viewpoint either, that is, while there may be fairly widespread agreement on the shortcomings of existing modelling procedures, there is considerable disagreement on the predicative power or value of these models.

There are a number of reasons to question the predictions of these models. Bryson (1989) suggests that the general circulation models are not wrong but just incomplete, in that they extrapolate results from the 30% of the land surface of the earth to the 70% occupied by water. As a result the models are accurate to about the same degree of temperature variation as between the late ice age and the present time. Moreover, the effects of volcanic aerosols and cloud are only incompletely considered, the end result being that models purport to explain small variations in temperature while they are unable to predict the so called "natural variation" in temperature. His basic conclusion seems to be that these relatively crude (although still very sophisticated) models are being pushed too far, past their degree of accuracy, in order to make long term predictions.

The economic effects of global warming hinge directly on the nature of the predictions from the atmospheric models. To the extent that uncertainty exists concerning the timing and degree of global warming, these uncertainties are then translated directly into the economic analysis. The relationship between the economic uncertainties, climate uncertainties, and policy uncertainties can be demonstrated by the following circular flow diagram. The following diagram demonstrates the central role of mining in this analysis.

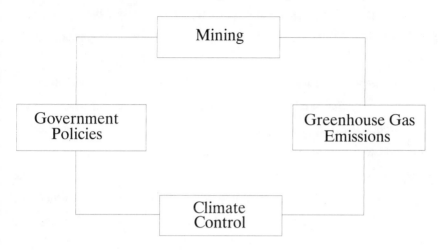

It is difficult to speak of an "optimal" greenhouse policy on the basis of the greatly varying scenarios, especially in light of the uncertainty which clouds the entire analysis including the impact on the mining sector. Nonetheless, some fairly general conclusions concerning the mining industry which focus on the policy link in the circular flow can still be drawn. First, the direct effects on the mining industry as a result of global warming appear to be minimal. Of more relevance are the indirect, induced, or secondary effects. These effects are felt as a result not of global warming itself, but as a consequence of policies intended to curb emissions of CO_2 and other greenhouse gases. These policies may have direct implications for the mining industry in terms of much higher energy prices as the result of carbon taxes, emissions taxes, coal export limitations, etc.

In this regard, a number of observations are relevant. First, recall that at least modest reductions in effective long wave absorbers can be made fairly inexpensively through limitations on CFC emissions. Subsequent reductions in GHG through carbon taxes and use taxes on fossil fuels are much less cost effective. Moreover, unilateral policy actions by the U.S. are suggested to have little effect on global warming even for energy taxes as high as 50%. The implications for the mining industry under the various policy options are not quite clear. Some suggest a resurgence in nuclear power, but that must be questioned on the basis of the obstacles faced by nuclear development over the last 15 years of so.[9] In terms of energy taxes it is difficult to assess the political viability of policies that appear to have little affect on global CO_2 concentrations.

The possible effects on the mining industry can be summarized in the following way. At the present time it would seem that various policies intended to limit carbon emissions have only limited effectiveness in the long run unless some means of curtailing growth in the energy industry is found. Given the ineffectiveness of these policies it becomes difficult to speculate on the likelihood of such policies being enacted, at least in the near future. Moreover, the really serious problems in terms of increased emissions appear to be delayed until the second quarter of the next century so that at the worst there seems to be a period of grace of 30-40 years in which the mining industry has to adapt to possible changes which might be forthcoming.

A comprehensive greenhouse policy is fraught with difficulties arising from the joint problem of regional variation of effects and separate jurisdictional entities. That is, while it would be difficult to envision a global carbon tax being imposed if all entities would benefit from this tax, if in fact some entities might benefit more in the absence of the tax, consensus and, hence, enforcement of such a policy would seem impossible. Nonetheless, the low cost potential benefits associated with reduced CFC emissions suggest that this might be at least an interim policy worth pursuing, especially with the added benefit of reducing the depletion of stratospheric ozone. This would at least be a temporary stopgap measure which could be employed to limit emissions while atmospheric models were being improved to the extent of determining whether or not the greenhouse effect was a phenomenon of extreme global concern. This policy arises as the natural extension of the fact that while there is not general agreement as to the nature, extent, and timing of the proposed greenhouse effect, there does seem to be a fairly widespread consensus that while we do not understand the problem completely, it would be both naive and myopic to assume it away.

NOTES

1.   The principle categories of chlorofluorocarbons contributing to global warming are the chlorofluoromethanes CFC11 and CFC12. These are used primarily as propellants and in refrigeration.

2.   For example, one molecule of CFC12 is 10^4 times more effective in trapping long-wave radiation than one molecule of CO_2 in the present atmosphere. (Mitchell, 1989)

3.   Due to the contribution of CFC11 and CFC12 to the depletion of stratoshperic ozone, actions have been taken to effectively eliminate these gases by the year 2000.

4.   70 degrees N is above the Arctic Circle, with the most severe effects predicted for the Queen Elizabeth Islands of Canada and northeastern Siberia.

5.   By known, we mean the ability of the trace gases to absorb the long-wave radiation and the associated increase in temperature, not the exact increase in the mean global temperature or the associated distributional and variational issues of temperature or the associated distributional and variational issues of temperature change.

6.   Direct and indirect effects, as we are using the terms here, are generally classified as market and non-market goods in the economics literature.

7.   Although the overall percentage will be less, it is assumed that the total global energy use will be greater.

8.   These calculations are based on the equation $A(t) = A(0) + .471 (f/r) (e^r - 1)C$ where $A(t)$ is atmospheric concentrations of CO^2 at time t, $A(0)$ is initial concentrations, f is the fraction of emissions that are airborne (not returning to the biomass), C is the rate of emissions in the given year, and r is the rate of growth of emissions.

9.   The basic question of trading environmental problems must also be considered, e.g., hazardous waste disposal v. CO_2 emissions.

REFERENCES

Bryson, R.A., 1989. "Will There be a Global 'Greenhouse' Warming?", Environmental Conservation, v. 16(2), pg. 97-9.

d'Arge, R., W.D. Schulze & D. Brookshire, 1982. "Carbon Dioxide and Intergenerational Choice", American Economic Review, v. 72(2), pg.251-6.

Edmonds, J. & J. Reilly, 1989. "Global Energy and CO_2 to the Year 2050", The Energy Journal, v. 4(3), pg. 21-47.

Hafele, W., 1981. Energy in a Finite World. Cambridge, MA: Ballinger.

Kellogg, W. & R. Schware, 1981. Climate Change and Society: Consequences of Increasing Atmospheric Carbon Dioxide. Boulder, CO: Westview Press.

Lave, L.B., 1982. "Mitigating Strategies for Carbon Dioxide Problems", American Economic Review, v. 72(2), pp. 257-61.

Mitchell, J.F.B., 1989. "The 'Greenhouse' Effect and Climate Change", Reviews of Geophysics, v. 27(1), pp. 115-39.

Nordhaus, W.D., 1977. "Economic Growth and Climate: The Carbon Dioxide Problem", American Economic Review, v. 67(1), pp. 341-6.

_____, 1982. "How Fast Should We Graze the Global Commons?", American Economic Review, v. 72(2), pp. 242-6.

_____, 1989. "The Economics of the Greenhouse Effect", Preliminary Paper, Yale University.

Olson, M., 1982. "Environmental Indivisibilities and Information Costs: Fanaticism, Agnosticism, and Intellectual Progress", American Economic Review, v. 72(2), pp. 262-6.

Ramanathan, V., et al., 1987. "Climate-Chemical Interactions and Effects of Changing Atmospheric Trace Gases", Reviews of Geophysics, v. 25(7), pp. 1441-82.

Resources for the Future, 1988. Greenhouse Warming: Abatement and Adaptation, Proceedings of a Workshop, June 14-15, Washington, D.C.

U.S. Environmental Protection Agency, 1989. "The Greenhouse Effect: How It Can Change Our Lives", EPA Journal, v. 15(1).

U.S. House of Representatives, "Technologies for Remediating Global Warming", Hearing before the Subcommittee of the Committee on Science, Space and Technology, June 29, 1988.

_____, "Energy Policy Implications of Global Warming", Hearings before the Committee on Energy & Commerce, July 7 and September 22, 1988.

U.S. Senate, "Greenhouse & Global Climate Change", Hearings before the Committee on Energy and Natural Resources, November 9 & 10, 1987.

Wood, W.B., G.J. Demko & P. Mofson, 1989. "Ecopolotics in the Global Greenhouse", Environment, v. 31(7).